Communications
in Computer and Information Science   436

T0212953

More information about this series at http://www.springer.com/series/7899

Dmitry I. Ignatov · Mikhail Yu. Khachay
Alexander Panchenko · Natalia Konstantinova
Rostislav E. Yavorskiy (Eds.)

Analysis of Images, Social Networks and Texts

Third International Conference, AIST 2014
Yekaterinburg, Russia, April 10–12, 2014
Revised Selected Papers

 Springer

Editors
Dmitry I. Ignatov
Rostislav E. Yavorskiy
National Research University Higher
  School of Economics
Moscow
Russia

Mikhail Yu. Khachay
Krasovsky Institute of Mathematics
  and Mechanics
Yekaterinburg
Russia

Alexander Panchenko
Université catholique de Louvain
Louvain-la-Neuve
Belgium

Natalia Konstantinova
University of Wolverhampton
Wolverhampton
UK

ISSN 1865-0929   ISSN 1865-0937 (electronic)
ISBN 978-3-319-12579-4   ISBN 978-3-319-12580-0 (eBook)
DOI 10.1007/978-3-319-12580-0

Library of Congress Control Number: 2014954576

Springer Cham Heidelberg New York Dordrecht London

Printed on acid-free paper

Springer is part of Springer Science+Business Media (www.springer.com)

Preface

This volume contains proceedings of the third conference on Analysis of Images, Social Networks, and Texts (AIST 2014). The first two conferences in 2012 and 2013 attracted a significant number of students, researchers, academics and engineers working on data analysis of images, texts, and social networks.

The broad scope of AIST makes it an event where researchers from different domains, such as image and text processing, exploiting various data analysis techniques, can meet and exchange ideas. We strongly believe that this may lead to cross-fertilisation of ideas between researchers relying on modern data analysis machinery. Therefore, AIST brings together all kinds of applications of data mining and machine learning techniques. The conference allows specialists from different fields to meet each other, present their work, and discuss both theoretical and practical aspects of their data analysis problems. Another important aim of the conference is to stimulate scientists and people from the industry to benefit from the knowledge exchange and identify possible grounds for fruitful collaboration.

Following an already established tradition, the conference was in Yekaterinburg, capital of Urals region in Russia during April 10–12, 2014. The key topics of AIST are analysis of images and videos; natural language processing and computational linguistics; social network analysis; machine learning and data mining; recommender systems and collaborative technologies; semantic web, ontologies and their applications; analysis of socioeconomic data.

The Program Committee and reviewers of the conference featured well-known experts in data mining and machine learning, natural language processing, image processing, and related areas from Russia and all over the world.

We received 74 high-quality submissions mostly from Russia but also from France, Germany, India, Poland, Spain, Ukraine, and USA, among which only 22 papers were accepted as regular oral papers (12 long and 10 short). Thus, the acceptance rate of this volume was around 30 %. In order to encourage young practitioners and researchers we included 3 short industry papers in the main volume and 33 submissions as posters in the supplementary proceedings. Each submission was reviewed by at least three reviewers, who are experts in their field, in order to supply detailed and helpful comments.

The conference also featured several invited talks and tutorials, as well as an industry session dedicated to current trends and challenges.

Invited talks:

- Boris Mirkin (Higher School of Economics, Moscow, Russia), Data Clustering: Some Topics of Current Interest
- Jaume Baixeries (Universitat Politècnica de Catalunya, Barcelona, Spain), Characterization of Database Dependencies with FCA and Pattern Structures

- Dmitriy Kolesov (NextGIS, Moscow, Russia), GIS as an Environment for Integration and Analysis of Spatial Data
- Natalia Konstantinova (University of Wolverhampton, UK), Relation Extraction – Let's Find More Knowledge Automatically

Tutorials:

- Jaume Baixeries (Universitat Politècnica de Catalunya, Barcelona, Spain), Introduction to Formal Concept Analysis and Attribute Dependencies (in 2 parts)
- Konstantin Voronstov (CCAS of RAS and Yandex, Moscow, Russia), Tutorial on Probabilistic Topic Modeling: Additive Regularization for Stochastic Matrix Factorization
- Natalia Konstantinova (University of Wolverhampton, UK), Introduction to Dialogue Systems, Personal Assistants are Becoming a Reality
- Natalia Konstantinova (University of Wolverhampton, UK), Academic Writing – Getting Published in International Journals and Conferences

The industry speakers also covered a rich variety of topics:

- Iosif Itkin (Exactpro Systems), Network Models for Exchange Trade Analysis
- Leonid I. Levkovitch-Maslyuk (EMC), Big Data for Business
- Alexander Semenov (http://jarens.ru/), Recent Advances in Social Network Analysis
- Irina Radchenko (http://iRadche.ru), Current Trends in Open Data and Data Journalism
- Oleg Lavrov (KM Alliance Russia), Knowledge Management as a Link between Business & IT
- Yury Kupriyanov (WikiVote!), IT Trends and Challenges in Knowledge Management

We would like to mention the best conference paper selected by the Program Committee. It was written by V.B. Surya Prasath and Radhakrishnan Delhibabu and entitled "Automatic Contrast Parameter Estimation in Anisotropic Diffusion for Image Restoration."

We would like to thank the authors for submitting their papers and to members of the Program Committee for their efforts in providing exhaustive reviews. We would also like to express special gratitude to all the invited speakers and industry representatives. We deeply thank all the partners and sponsors, and owe our gratitude to the Scientific Fund of Higher School of Economics for providing five AIST participants with travel grants. Our special thanks go to Springer editors who helped us, starting from the first conference call to the final version of the proceedings. Last but not least, we are grateful to all organisers, especially to Eugeniya Vlasova and Dmitry Ustalov, whose endless energy saved us in the most critical stages of the conference preparation.

The Russian word "aist" is more than just a simple abbreviation ("аист" in Cyrillic), it means a "stork". Since it is a wonderful free bird, a symbol of happiness and peace, this stork brought us the inspiration to organise the AIST conference. Storks, which can nest even in the polar areas of Russia, make long trips to South Europe, Middle East, India, and North Africa every year. So we believe that this young and rapidly growing conference will be bringing inspiration to data scientists around the World!

May 2014

Dmitry I. Ignatov
Mikhail Yu. Khachay
Alexander Panchenko
Natalia Konstantinova
Rostislav E. Yavorskiy

Organisation

The conference was organized by a joint team from Krasovsky Institute of Mathematics and Mechanics, Ural Branch of Russian Academy of Sciences (Yekaterinburg, Russia) and the National Research University Higher School of Economics (Moscow, Russia). It was supported by the Russian Foundation for Basic Research Grant No. 14-07-06803.

Program Committee Chairs

Mikhail Yu. Khachay | Krasovsky Institute of Mathematics and Mechanics of UB RAS, Russia
Dmitry I. Ignatov | National Research University Higher School of Economics, Russia
Alexander Panchenko | Université catholique de Louvain, Belgium

Organising Chair

Rostislav E. Yavorskiy | National Research University Higher School of Economics, Russia

Proceedings Chair

Natalia Konstantinova | University of Wolverhampton, UK

Poster Chair

Dmitry Ustalov | Krasovsky Institute of Mathematics and Mechanics and Ural Federal University, Russia

Organising Committe

Eugeniya Vlasova | National Research University Higher School of Economics, Moscow, Russia
Dmitry Ustalov | Krasovsky Institute of Mathematics and Mechanics and Ural Federal University, Yekaterinburg, Russia
Olga Fazylova | National Research University Higher School of Economics, Moscow, Russia
Liliya Galimziyanova | Ural Federal University, Yekaterinburg, Russia

Alexandra Kaminskaya	Yandex, Moscow, Russia
Maria Khakimova	Perm State University, Perm, Russia
Natalia Korepanova	National Research University Higher School of Economics, Moscow, Russia
Elena Nikulina	National Research University Higher School of Economics, Moscow, Russia
Andrey Shelomentsev	APIO, Yekaterinburg and Moscow, Russia
Anna Voronova	Yandex, Moscow, Russia

Program Committee

Tinku Acharya	Videonetics Technology Pvt. Ltd., India
Simon Andrews	Sheffield Hallam University, UK
Olga Barinova	Moscow State University and Yandex, Russia
Malay Bhattacharyya	University of Kalyani, India
Vladimir Bobrikov	Imhonet Research, Russia
Elena Bolshakova	Moscow State University and National Research University Higher School of Economics, Russia
Jean-Leon Bouraoui	Prometil SARL, France
Pavel Braslavski	Ural Federal University and Kontur Labs, Russia
Ekaterina Chernyak	National Research University Higher School of Economics, Russia
Ilia Chetviorkin	Moscow State University, Russia
Marina Chicheva	Image Processing Systems Institute of RAS, Russia
Florent Domenach	University of Nicosia, Cyprus
Alexey Drutsa	Moscow State University and Yandex, Russia
Maxim Dubinin	NextGIS, Russia
Victor Erukhimov	Itseez, Russia
Rashid Faizullin	Omsk State Technical University, Russia
Thomas Francois	Université catholique de Louvain, Belgium
Boris Galitsky	Become.com, USA
Dmitry Ilvovsky	Fors and National Research University Higher School of Economics, Russia
Vladimir Ivanov	Kazan Federal University, Russia
Vadim Kantorov	Inria, France
Mehdi Kaytoue	LIRIS – INSA de Lyon, France
Vladimir Khoroshevsky	Dorodnicyn Computing Centre of RAS and National Research University Higher School of Economics, Russia
Sergei Koltsov	National Research University Higher School of Economics, Russia
Olessia Koltsova	National Research University Higher School of Economics, Russia
Natalia Konstantinova	University of Wolverhampton, UK
Anton Konushin	Lomonosov Moscow State University, Russia

Konstantin Vorontsov	Dorodnicyn Computing Centre of RAS, Russia
Leonind Zhukov	Ancestry.com, USA and National Research University Higher School of Economics, Russia
Nataly Zhukova	Saint Petersburg institute for Infromatics and Automation of RAS, Russia
Dominik Ślęzak	University of Warsaw, Poland and Infobright Inc., Canada

Invited Reviewers

Aleksei Buzmakov	Inria, France
Yuri Katkov	Blue Brain Project, Switzerland
Lidia Pivovarova	University of Helsinki, Finland
Alexander Semenov	National Research University Higher School of Economics, Russia
Nikita Spirin	University of Illinois at Urbana-Champaign, USA
Dmitry Ustalov	Krasovsky Institute of Mathematics and Mechanics and Ural Federal University, Russia
Andrey Bronevich	National Research University Higher School of Economics, Russia
Sujoy Chatterjee	University of Kalyani, India
Surya Prasath	University of Missouri-Columbia, USA
Kirill Shileev	Digital Society Laboratory, Russia
Natalia Vassilieva	HP Labs, Russia
Victoria Yaneva	University of Wolverhampton, UK

Partners and Sponsoring Institutions

Russian Foundation for Basic Research
Yandex
CLAIM
Data Mining Labs
Digital Society Laboratory
EMC
GraphiCon
Innovative Trading Systems
Krasovsky Institute of Mathematics and Mechanics
NextGIS
NLPub
penxy
SEC(R)
SKB Kontur
Ural Federal University
Ural IT Cluster
WikiVote!

Contents

Industry Papers

Invited Papers and Tutorials

Characterization of Database Dependencies with FCA and Pattern Structures

Jaume Baixeries[1]([✉]), Mehdi Kaytoue[2], and Amedeo Napoli[3]

[1] Universitat Politècnica de Catalunya, 08032 Barcelona, Catalonia, Spain
jbaixer@lsi.upc.edu
[2] Université de Lyon, CNRS, INSA-Lyon, LIRIS, UMR5205, Lyon 69621, France
mehdi.kaytoue@insa-lyon.fr
[3] LORIA (CNRS - Inria Nancy Grand Est - Université de Lorraine), B.P. 239,
54506 Vandœuvre-lès-Nancy, France
amedeo.napoli@loria.fr

Abstract. In this review paper, we present some recent results on the characterization of Functional Dependencies and variations with the formalism of Pattern Structures and Formal Concept Analysis.

Although these dependencies have been paramount in database theory, they have been used in different fields: artificial intelligence and knowledge discovery, among others.

Keywords: Attribute implications · Data dependencies · Pattern structures · Formal concept analysis · Data analysis

1 Database Dependencies in Data Analysis

In the relational Database Model, a database dependency describes a relationship between sets of attributes that hold in a table. To give a simple and intuitive example of such relations, les us suppose that we have a table with personal details (technically: attributes) of different people: name, age, birth place, average income, and so on. A dependency could state, for instance, that *the age of a person is related to his average income*. This notion of relationship between the attributes of the table is the central idea of a **Database Dependency**. Obviously, this generic idea of relationship may hold different sintactical and semantical formalizations.

However, the idea of data dependency is not only confined to the realm of Database Theory. It can be found in different domains such as Artificial Intelligence, logics or Formal Concept Analysis (FCA), to name a few.

In Artificial Intelligence we may mention, as an example, Decision Trees, which induce a set of relations between the nodes in that tree. In logics, we may mention Horn clauses that formalize the dependency that exist between variables [2]. In FCA we have implications, which explain the dependencies that exist between the attributes in a context [12].

In this paper, we focus on database dependencies that have been defined in the relational database model. In fact, we will focus on functional dependencies (FDs) and derivate dependencies. These are among the most popular

© Springer International Publishing Switzerland 2014
D.I. Ignatov et al. (Eds.): AIST 2014, CCIS 436, pp. 3–14, 2014.
DOI: 10.1007/978-3-319-12580-0_1

types of dependencies [31] since they indicate a functional relation between sets of attributes: the values of a set of attributes are determined by the values of another set of attributes. To handle errors and uncertainty in real-world data, alternatives exist. *Approximate Dependencies* [15] are FDs that hold in a part –which is user defined– of the database. *Purity Dependencies* [27] express the relationship on the relative impurity induced by two partitions of the table (generated by two sets of attributes). If the impurity is zero, we have a FD.

Another alternative are *Similarity Dependencies* [5], which can be seen as a generalization of Functional Dependencies, but un-crispring the basic definition of FDs: similar values of an attribute determine similar values of another attribute. Similarity has been considered for FDs under several terms, e.g. fuzzy FDs [8], matching dependencies [29], constraint generating dependencies [7]. Moreover, it is still an active topic of research in the database community [9,10,29,30].

Because of the interest that those dependencies have in Data Analysis, in this paper we want to address their characterization with the formalism of Formal Concept Analysis [12]. The choice of this formalism is explained by the important research in this field that has been devoted to the computation of implications, functional dependencies, multivalued dependencies, and other kinds of dependencies [1,3–5,8,21–23,34]. Formal Concept Analysis has been proved, as well, as a valid formalism for different fields of Knowledge Discovery [20,24,25], Machine Learning [18] or gene mining [16,17], among others.

This paper is organized as follows: first, we present the formal definitions for the different kinds of dependencies that will be discussed, focusing in detail in Functional and Similarity Dependencies. Next, we present some computational issues of those dependencies within the classical FCA framework, and, finally, we present an alternative framework. The main goal of this paper is to present recent advancements in the characterization of functional and similarity dependencies, so that they can be easily used in the process of analyzing large datasets.

2   Functional Dependencies and Variations

2.1   Notation

We deal with datasets which are sets of tuples. Let $\mathcal{U}$ be a set of attributes and Dom be a set of values (a domain). For the sake of simplicity, we assume that Dom is a numerical set. A tuple t is a function $t : \mathcal{U} \mapsto Dom$ and then a table T is a set of tuples. Usually a table is presented as a matrix, as in the table of Example 1, where the set of tuples (or objects) is $T = \{t_1, t_2, t_3, t_4\}$ and $\mathcal{U} = \{a, b, c, d\}$ is the set of attributes.

The functional notation allows one to associate an attribute with its value. We define the functional notation of a tuple for a set of attributes X as follows, assuming that there exists a total ordering on $\mathcal{U}$. Given a tuple $t \in T$ and $X = \{x_1, x_2, \ldots, x_n\} \subseteq \mathcal{U}$, we have:

$$t(X) = \langle t(x_1), t(x_2), \ldots, t(x_n) \rangle$$

In Example 1, we have $t_2(\{a,c\}) = \langle t_2(a), t_2(c) \rangle = \langle 4, 4 \rangle$. In this paper, the set notation is usually omitted and we write ab instead of $\{a,b\}$.

Example 1. This is an example of a table $T = \{t_1, t_2, t_3, t_4\}$, based on the set of attributes $\mathcal{U} = \{a, b, c, d\}$.

id	a	b	c	d
t_1	1	3	4	1
t_2	4	3	4	3
t_3	1	8	4	1
t_4	4	3	7	3

We are also dealing with the set of partitions of a set. Let S be any arbitrary finite set, then $\mathrm{Part}(S)$ is the set of all possible partitions that can be formed with S. The set of partitions of a set is a lattice [13]. We recall that partitions can also be considered as equivalence classes induced by an equivalence relation.

2.2 Functional Dependencies

We now introduce functional dependencies (FDs).

Definition 1 [31]. *Let T be a set of tuples (or a data table), and $X, Y \subseteq \mathcal{U}$. A functional dependency (FD) $X \rightarrow Y$ holds in T if*

$$\forall t, t' \in T : t(X) = t'(X) \implies t(Y) = t'(Y)$$

For example, the functional dependencies $a \rightarrow d$ and $d \rightarrow a$ hold in the table of Example 1, whereas the functional dependency $a \rightarrow c$ does not hold since $t_2(a) = t_4(a)$ but $t_2(c) \neq t_4(c)$.

There is an alternative way of considering Functional Dependencies using partitions of the set of tuples T. Taking a set of attributes $X \subseteq \mathcal{U}$, we define the partition of tuples induced by this set as follows.

Definition 2. *Let $X \subseteq \mathcal{U}$ be a set of attributes in a table T. Two tuples t_i and t_j in T are equivalent w.r.t. X when:*

$$t_i \sim t_j \iff t_i(X) = t_j(X)$$

Then, the partition of T induced by X is a set of equivalence classes:

$$\Pi_X(T) = \{c_1, c_2, \ldots, c_m\}$$

For example, if we consider the table in Example 1, we have $\Pi_a(T) = \{\{t_1, t_3\}, \{t_2, t_4\}\}$.

The set of all partitions of a set T is $\mathrm{Part}(T)$. We can also notice that the set of partitions of any set $\mathrm{Part}(T)$ induces an ordering relation $\leq$:

$$\forall P_i, P_j \in \mathrm{Part}(T) : P_i \leq P_j \iff \forall c \in P_i : \exists c' \in P_j : c \subseteq c'$$

For example: $\{\{t_1\}, \{t_2\}, \{t_3, t_4\}\} \leq \{\{t_1\}, \{t_2, t_3, t_4\}\}$. According to the partitions induced by a set of attributes, we have an alternative way of defining necessary and sufficient conditions for a functional dependency to hold:

Proposition 1 [15]. *A functional dependency $X \to Y$ holds in T if and only if $\Pi_Y(T) \leq \Pi_X(T)$.*

Again, taking the table in Example 1, we have that $a \to d$ holds and that $\Pi_d \leq \Pi_a$ since $\Pi_a(T) = \{\{t_1, t_3\}, \{t_2, t_4\}\}$ and $\Pi_d(T) = \{\{t_1, t_3\}, \{t_2, t_4\}\}$ (actually $d \to a$ holds too).

2.3 Purity and Approximate Dependencies

We now present the definition of two generalizations of Functional Dependencies: Purity Dependencies and Approximate Dependencies.

Approximate Dependencies [15]. In a table, there may be some tuples that prevent a functional dependency from holding. Those tuples can be seen as exceptions (or errors) for that dependency. Removing such tuples allows the dependency to exist: then a threshold can be set to define a set of "approximate dependencies" holding in a table. For example, a threshold of 10 % means that all functional dependencies holding after removing up to 10 % of the tuples of a table are valid approximate dependencies. The set of tuples to be removed for validating a functional dependency does not need to be the same for

Example 2. This table is an excerpt of the *Average Daily Temperature Archive*[1] from The University of Dayton, that shows the month average temperatures for different cities.

id	Month	Year	Av. Temp.	City
t_1	1	1995	36.4	Milan
t_2	1	1996	33.8	Milan
t_3	5	1996	63.1	Rome
t_4	5	1997	59.6	Rome
t_5	1	1998	41.4	Dallas
t_6	1	1999	46.8	Dallas
t_7	5	1996	84.5	Houston
t_8	5	1998	80.2	Houston

each approximate dependency. Considering the dependency in Example 2 *Month* $\to$ *Av.Temp*, we can check that 6 tuples should be removed before verifying the dependency: we keep only one tuple for Month 1 and one tuple for Month 5 (actually just as if we remove "duplicates"). Then, if the threshold is equal to or larger than 75 %, *Month* $\to$ *Av.Temp* is a valid Approximate Dependency.

Purity Dependencies [27] are a generalization of the relationship between partitions induced by the left-hand side and right-hand side of a functional dependency. These dependencies are based on the relative impurity measure of two partitions. In order to compute this impurity measure, we need a concave and subadditive function defined on the interval $[0, 1]$ (for example, the binary entropy function). The intuition about this measure is that it computes *how much those partitions disagree*, i.e. how far two partitions π and σ are from fulfilling the relation $\pi \leq \sigma$. In the case of approximate dependencies, this measure

---

[1] http://academic.udayton.edu/kissock/http/Weather/

was the number (or percentage) of tuples that had to be removed, but as for purity dependencies, this measure can be more general. If the impurity measure is zero (or close to zero), then $\pi \leq \sigma$. It indicates that those tuples fulfil (or almost fulfil) the relation $\pi \leq \sigma$. The *closer* they are to the relation $\pi \leq \sigma$, the closer to zero will be their relative impurity.

For example, the impurity measure (details on this measure are given in [26]) of partition $\{\{1,2,3\},\{4,5\}\}$ w.r.t. partition $\{\{1,2\},\{3,4,5\}\}$ is 5.6, whereas the impurity measure of partition $\{\{1,3\},\{2,5\},\{4\}\}$ w.r.t. partition $\{\{1,2\},\{3,4,5\}\}$ is 8.2. In the first pair of partitions, only tuple 3 is misplaced, i.e. moving 3 from one partition to another leads to the same partitions, whereas in the second example, the number of misplaced elements is larger (2, 3, and 4 should be moved).

An important feature of this measure is that if a partition is finer than another, then, their relative impurity measure is exactly 0. This implies that a purity dependency $X \rightarrow Y$ holds if and only if the relative impurity of $\Pi_X(T)$ w.r.t. $\Pi_Y(T)$ is below a user-defined threshold. Therefore, if $\Pi_Y(T) \leq \Pi_X(T)$, a functional dependency is a valid purity dependency, regardless of the threshold.

2.4 Similarity Dependencies

Similarity Dependencies [5] (these dependencies are called Differential Dependencies in [28]) are another generalization of Funcional Dependencies. The intuition behind these dependencies is that in Similarity Dependencies, we relax the condition that states that $X \rightarrow Y$ holds if and only if for each pair of tuples, the fact that if their values in attributes X are the same, their values in Y are the same as well. For similarity dependencies, we change the *are the same* by *are similar*.

In order to define Similarity Dependencies, we first define a *tolerance* relation in a set S:

Definition 3. $\theta \subseteq S \times S$ *is a tolerance relation if:*

1. $\forall s_i \in S : s_i \theta s_i$ *(reflexivity)*
2. $\forall s_i, s_j \in S : s_i \theta s_j \iff s_j \theta s_i$ *(symmetry)*

A tolerance relation is not necessarily transitive. Precisely, in this detail lies the difference between Functional and Similarity Dependencies: in FD's we had that the relation $=$ induced an equivalence relation, because it is obviously transitive, but in this case, we do not need to enforce this condition.

An example of a tolerance relation is the *similarity* that can be defined within a set of integer values as follows. Given two integer values v_1, v_2 and a threshold ϵ (user-defined): $v_1 \theta v_2 \iff |v_1 - v_2| \leq \epsilon$. For example, when $S = \{1,2,3,4,5\}$ and $\epsilon = 2$, then $S/\theta = \{\{1,2,3\},\{2,3,4\},\{3,4,5\}\}$. S/θ is not a partition as transitivity does not apply.

A tolerance relation induces *blocks of tolerance*:

Definition 4. *Given a set S, a subset $K \subseteq S$, and a tolerance relation $\theta \subseteq S \times S$, K is a* block of tolerance *of θ if:*

1. $\forall x, y \in K : x\theta y$ *(pairwise correspondence)*
2. $\forall z \notin K, \exists u \in K : \neg(z\theta u)$ *(maximality)*

All elements in a tolerance block are in pairwise correspondence, and the block is maximal with respect to the relation θ. The set of all tolerance blocks induced by a tolerance relation θ on the set S is denoted by S/θ (by analogy with the notation of equivalence classes). S/θ is a set of maximal subsets of S and as such, $S/\theta \in \wp(\wp(S))$. The set of sets of tolerance blocks also induces a partial order, as in the case of partitions. In fact, a tolerance relation and tolerance blocks are a generalization of equivalence relations and equivalence classes.

Two tuples are similar w.r.t. a set of attributes X if and only if they are similar w.r.t. *each* attributes in X. We now can define a *similarity dependency*:

Definition 5. *Let $X, Y \subseteq \mathcal{U}$: $X \to Y$ is a similarity dependency iff $\forall t_i, t_j \in T$: $t_i\theta_X t_j \implies t_i\theta_Y t_j$*

Example 3. Going back to Example 2, let us compute the Similarity Dependencies that hold and that have the attribute *Av. Temp.* in their right-hand side).

Dependency	Holds
Month -> Av. Temp	N
Month, Year -> Av. Temp	N
Month, City -> Av. Temp	Y
Year -> Av. Temp	N
Year, City -> Av. Temp	N
City -> Av. Temp	N

The only similarity dependency that holds is $Month, City \to Av.Temp$, using the following similarity measures for each attribute: $x\ \theta_{Month}\ y \iff |x - y| \leq 0$, $x\ \theta_{Year}\ y \iff |x - y| \leq 0$, $x\ \theta_{City}\ y \iff distance(x, y) \leq 500$ and $x\ \theta_{Av.Temp}\ y \iff |x - y| \leq 10$.

The similarity imposes that the month and year must be the same, whereas the distance between cities should be less than 500 km and the difference between average temperatures should be less than 10 degrees (all these values are of course arbitrary).

In particular, considering the tuples t_1, t_2: $t_1\theta_{Month,City}t_2$ since $t_1(Month) = t_2(Month) = \langle 1 \rangle$ and $t_1(City) = t_2(City) = \langle Milan \rangle$. From the other side, we have that $t_1\theta_{Av.Temp.}t_2$ since $|36.4 - 33.8| \leq 10$.

3   Formal Concept Analysis

Formal Concept Analysis (FCA) [12] is a mathematical framework allowing to build a concept lattice from a binary relation between objects and their attributes. The concept lattice can be represented by a diagram where classes of objects/attributes and ordering relations between classes can be drawn, interpreted and used for data-mining, knowledge management and discovery [32,33].

We use standard definitions from [12]. Let G and M be arbitrary sets and $I \subseteq G \times M$ be an arbitrary binary relation between G and M. The triple (G, M, I) is called a formal context. Each $g \in G$ is interpreted as an object, each $m \in M$ is interpreted as an attribute. The statement $(g, m) \in I$ is interpreted as "g has attribute m". The two following derivation operators $(\cdot)'$:

$$A' = \{m \in M \mid \forall g \in A : gIm\} \qquad for\ A \subseteq G,$$
$$B' = \{g \in G \mid \forall m \in B : gIm\} \qquad for\ B \subseteq M$$

define a Galois connection between the powersets of G and M. The derivation operators $\{(\cdot)', (\cdot)'\}$ put in relation elements of the lattices $(\wp(G), \subseteq)$ of objects and $(\wp(M), \subseteq)$ of attributes and reciprocally. A Galois connection induces closure operators $(\cdot)''$ and realizes a one-to-one correspondence between all closed sets of objects and all closed sets of attributes. For $A \subseteq G$, $B \subseteq M$, a pair (A, B) such that $A' = B$ and $B' = A$, is called a *formal concept*. Concepts are partially ordered by $(A_1, B_1) \leq (A_2, B_2) \Leftrightarrow A_1 \subseteq A_2 (\Leftrightarrow B_2 \subseteq B_1)$. (A_1, B_1) is a sub-concept of (A_2, B_2), while the latter is a super-concept of (A_1, B_1). With respect to this partial order, the set of all formal concepts forms a complete lattice called the *concept lattice* of the formal context (G, M, I), i.e. any subset of concepts has both a supremum (join $\vee$) and an infimum (meet $\wedge$) [12]. For a concept (A, B) the set A is called the *extent* and the set B the *intent* of the concept. The set of all concepts of a formal context (G, M, I) is denoted by $\mathfrak{B}(G, M, I)$ while the concept lattice is denoted by $\underline{\mathfrak{B}}(G, M, I)$.

An implication of a formal context (G, M, I) is denoted by $X \rightarrow Y, X, Y \subseteq M$ and means that all objects from G having the attributes in X also have the attributes in Y, i.e. $X' \subseteq Y'$. Implications obey the Amstrong rules (reflexivity, augmentation, transitivity). A minimal subset of implications (in sense of its cardinality) from which all implications can be deduced with Amstrong rules is called the Duquenne-Guigues basis [14].

Objects described by non binary attributes can be represented in FCA as a many-valued context (G, M, W, I) with a set of objects G, a set of attributes M, a set of attribute values W and a ternary relation $I \subseteq G \times M \times W$. The statement $(g, m, w) \in I$, also written $g(m) = w$, means that "the value of attribute m taken by object g is w". The relation I verifies that $g(m) = w$ and $g(m) = v$ always implies $w = v$. For applying the FCA machinery, a many-valued context can be transformed into a formal context with a conceptual scaling. The choice of a scale should be wisely done w.r.t. data and goals since it affects the size, the interpretation, and the computation of the resulting concept lattice. This will be discussed in the next section.

4   Computation of FD's

There are different algorithms that deal with the computation of FD's. One of the most well-known is TANE [15], which is based on the incremental computation of equivalence classes of tuples. FD Mine [35] takes the approach of reducing the number of candidate dependencies by using the axioms that apply to them. Approximate Dependencies computation is explained in [15], and other different kind of FD-alike dependencies can be found in [29] and [30].

In this paper, we focus on the computation of FD's and Similarity Dependencies within the framework of Formal Concept Analysis. The most well-known method [1,12] is called *binarization*, and consists in transforming (implicitly) a many-valued set of data into a binary context. This transformation allows us to build a formal context. In order to define this formal context, we need to define first the set of objects:

$$G = \{(t_i, t_j) \mid i < j \text{ and } t_i, t_j \in T\}$$

This corresponds to the set of all pairs of tuples from T (excluding symmetry and reflexivity). The relation of the context is defined as:

$$(t_i, t_j)\ I\ x \Leftrightarrow t_i(x) = t_j(x)$$

It is important to realize that the formal context $\mathbb{K} = (G, \mathcal{U}, I)$ depends entirely on the table T, since both G and $\mathcal{U}$ do, but this dependency is not explicitly shown in the definition of this context. Figure 1 presents an example of this binarization.

We can see that the size of this context can be of the order of $\mathcal{O}(|T^2|)$ (where $|T|$ is the number of tuples of T), so it can be significantly bigger than the original set of data.

id	a	b	c	d
t_1	1	2	3	1
t_2	1	2	1	4
t_3	1	1	3	4
t_4	2	2	3	4

id	a	b	c	d
(t_1, t_2)	x	x		
(t_1, t_3)	x		x	
(t_1, t_4)			x	x
(t_2, t_3)	x			x
(t_2, t_4)		x		x
(t_3, t_4)			x	x

Fig. 1. A data table T (left) with its associated formal context $(\mathcal{B}_2(G), M, I)$ (right).

Another FCA-oriented option is that of **scaling**. This alternative consists in creating a formal context such that its attributes are all possible values that can appear in a table (see [12] for a more detailed explanation). The problem of this approach is that it depends on the number of different values that a table may contain, which, in most cases, can be an extremely large number of them.

Both methods suffer from the same drawback: the fact that, the transformation of the original data into a formal context can be too costly in terms of size w.r.t. the original dataset.

In order to overcome this burden, [4] and [5] propose the use of Pattern Structures (an extension of Formal Concept Analysis). In the next section, we review some details of this approach.

5 Pattern Structures in Formal Concept Analysis

Our interest lies in handling numerical data within FCA. Hence, we recall here the formalism of pattern structures that can be understood as a generalization towards complex data, i.e. objects taking descriptions in a partially ordered set. In the case of pattern structures, the descriptions will be taken from meet–semi-lattices, which arise naturally from a partial order [19].

A pattern structure is defined as a generalization of a formal context describing complex data [11]. Formally, let G be a set of objects, let $(D, \sqcap)$ be a meet-semi-lattice of potential object descriptions and let $\delta : G \longrightarrow D$ be a mapping associating each object with its description. Then $(G, (D, \sqcap), \delta)$ is a pattern structure. Elements of D are patterns and are ordered by a subsumption relation $\sqsubseteq$: $\forall c, d \in D, c \sqsubseteq d \Longleftrightarrow c \sqcap d = c$. A pattern structure $(G, (D, \sqcap), \delta)$ gives rise to two derivation operators $(\cdot)^\square$:

$$A^\square = \bigsqcap_{g \in A} \delta(g) \quad for \ A \subseteq G$$

$$d^\square = \{g \in G | d \sqsubseteq \delta(g)\} \quad for \ d \in (D, \sqcap).$$

These operators form a Galois connection between $(2^G, \subseteq)$ and $(D, \sqcap)$. Pattern concepts of $(G, (D, \sqcap), \delta)$ are pairs of the form (A, d), $A \subseteq G$, $d \in (D, \sqcap)$, such that $A^\square = d$ and $A = d^\square$. For a pattern concept (A, d), d is a pattern intent and is the common description of all objects in A, the pattern extent. When partially ordered by $(A_1, d_1) \leq (A_2, d_2) \Leftrightarrow A_1 \subseteq A_2 \ (\Leftrightarrow d_2 \sqsubseteq d_1)$, the set of all concepts forms a complete lattice called pattern concept lattice.

6 Functional Dependencies and Their Variations with FCA and Pattern Structures

Consider a numerical table as a many-valued context (G, M, W, I) where G corresponds to the set of objects ("rows"), M to the set of attributes ("columns"), W the data domain ("all distinct values of the table") and $I \subseteq G \times M \times W$ a relation such that $(g, m, w) \in I$ also written $m(g) = w$ means that attribute m takes the value w for the object g [12].

We now show how the Functional Dependencies and Similarity Dependencies can be characterized using pattern structures and FCA. We first have to define the set of formal objects, which in both cases are the set of attributes that are present in the original table. Then, given an attribute $m \in M$, its description $\delta(m)$ is given by the sets induced by the respective relations within the set of tuples: the equivalence relation defined in Sect. 2.2 for Functional Dependencies and the tolerance relation defined in Sect. 2.4 for Similarity Dependencies. In both cases, the description of an attribute is a set of sets of tuples. In the case of FD's, this will be a partition, in the case of Similarity Dependencies, it will be a set of tolerance blocks. For instance, in the case of Functional Dependencies, the description of an attribute will be given by a partition over G such that any two elements g, h of the same class take the same values for the attribute m, i.e. $m(g) = m(h)$.

Since in both cases the set of descriptions obey to a partial order, our initial numerical table (G, M, W, I) can be represented as a pattern structure $(M, (D, \sqcap, \sqcup), \delta)$ where M is the set of original attributes, and $(D, \sqcap, \sqcup)$ is the lattice of partitions or tolerance blocks over G.

Therefore, we have that, for a given set of attributes $X \subseteq \mathcal{U}$, its description is $\{X\}^{\square}$. We remark that $\{X\}^{\square}$ depends on the definition of the pattern formal context defined, which, in turn, depends on the definition of the description of the attributes. Table 1 shows an example of the description of all the attributes of a table, in the case that the description is computed according to the binary relation in Definition 2.

Table 1. The original data (left), the resulting pattern structure (right)

id	A	B	C	D
1	1	3	7	2
2	1	3	4	5
3	3	5	2	2
4	3	3	4	8

$m \in M$	$\delta(m) \in (D, \sqcap, \sqcup)$
A	$\{\{1,2\}, \{3,4\}\}$
B	$\{\{1,2,4\}, \{3\}\}$
C	$\{\{1\}, \{2,4\}, \{3\}\}$
D	$\{\{1,3\}, \{2\}, \{4\}\}$

We now can state how this formal pattern context characterizes the set of Functional Dependencies or Similarity Dependencies that hold in a dataset:

Proposition 2 [5]. *A functional (similarity) dependency $X \to Y$ holds in a table T if and only if: $\{X\}^{\square} = \{XY\}^{\square}$ in the pattern structure $(M, (D, \sqcap), \delta)$.*

7 Conclusions and Future Work

In this paper we have presented Functional Dependencies and other generalizations, and we have discussed how those dependencies are relevant and of interest to data analysis, and we have focused on FCA-based characterizations.

We have also discussed that the classical methods in FCA for computing dependencies have a computational cost that may be, in some cases, unfeasable. Pattern Structures are a way to overcome this problem. We have presented a way to apply this framework in order to characterize Functional and Similarity Dependencies. Experiments [5,6] seem to confirm the validity of this approach.

Future work should advance into two different (yet, complementary) paths: on the one hand, it is needed to perform more experiments and evaluate them more precisely, in terms of speed as well as memory usage. On the other hand, it is also needed to use this same framework in order to characterize other kinds of dependencies which have been discussed in this paper. This would include Approximate, Purity, Fuzzy Dependencies, among others.

Acknowledgments. This research work has been supported by the Spanish Ministry of Education and Science (project TIN2008-06582-C03-01), EU PASCAL2 Network of Excellence, and by the Generalitat de Catalunya (2009-SGR-980 and 2009-SGR-1428) and AGAUR (grant 2010PIV00057).

References

1. Baixeries, J.: Lattice Characterization of Armstrong and Symmetric Dependencies (Ph.D. Thesis). Universitat Politècnica de Catalunya, (2007)
2. Baixeries, J., Balcázar, J.L.: Discrete deterministic data mining as knowledge compilation. In: Proceedings of Workshop on Discrete Mathematics and Data Mining - SIAM (2003)
3. Baixeries, J., Balcázar, J.L.: A lattice representation of relations, multivalued dependencies and armstrong relations. In: ICCS, pp. 13–26 (2005)
4. Baixeries, J., Kaytoue, M., Napoli, A.: Computing functional dependencies with pattern structures. In: Szathmary, L., Priss, U., (eds.) CLA. CEUR Workshop Proceedings, vol. 972, pp. 175–186. CEUR-WS.org (2012)
5. Baixeries, J., Kaytoue, M., Napoli, A.: Computing similarity dependencies with pattern structures. In: CLA, pp. 33–44 (2013)
6. Baixeries, J., Kaytoue, M., Napoli, A.: Characterizing functional dependencies in formal concept analysis with pattern structures. Ann. Math. Artif. Intell. **72**, 1–21 (2014)
7. Baudinet, M., Chomicki, J., Wolper, P.: Constraint-generating dependencies. J. Comput. Syst. Sci. **59**(1), 94–115 (1999)
8. Bělohlávek, R., Vychodil, V.: Data tables with similarity relations: functional dependencies, complete rules and non-redundant bases. In: Li Lee, M., Tan, K.-L., Wuwongse, V. (eds.) DASFAA 2006. LNCS, vol. 3882, pp. 644–658. Springer, Heidelberg(2006)
9. Bertossi, L., Kolahi, S., Lakshmanan, L.V.S.: Data cleaning and query answering with matching dependencies and matching functions. In: Proceedings of the 14th International Conference on Database Theory, ICDT '11, pp. 268–279. ACM, New York (2011)
10. Fan, W., Gao, H., Jia, X., Li, J., Ma, S.: Dynamic constraints for record matching. The VLDB J. **20**(4), 495–520 (2011)
11. Ganter, B., Kuznetsov, S.O.: Pattern structures and their projections. In: Delugach, H.S., Stumme, G. (eds.) ICCS 2001. LNCS (LNAI), vol. 2120, pp. 129–142. Springer, Heidelberg (2001)
12. Ganter, B., Wille, R.: Formal Concept Analysis. Springer, Berlin (1999)
13. Graetzer, G., Davey, B., Freese, R., Ganter, B., Greferath, M., Jipsen, P., Priestley, H., Rose, H., Schmidt, E., Schmidt, S., Wehrung, F., Wille, R.: General Lattice Theory. Freeman, San Francisco (1971)
14. Guigues, J.-L., Duquenne, V.: Familles minimales d'implications informatives résultant d'un tableau de données binaires. Mathématiques et Sciences Humaines **95**, 5–18 (1986)
15. Huhtala, Y., Kärkkäinen, J., Porkka, P., Toivonen, H.: Tane: an efficient algorithm for discovering functional and approximate dependencies. Comput. J. **42**(2), 100–111 (1999)
16. Kaytoue, M., Kuznetsov, S.O., Napoli, A.: Revisiting numerical pattern mining with formal concept analysis. In: IJCAI, pp. 1342–1347 (2011)

17. Kaytoue, M., Kuznetsov, S.O., Napoli, A., Duplessis, S.: Mining gene expression data with pattern structures in formal concept analysis. Inf. Sci. **181**(10), 1989–2001 (2011)

18. Kuznetsov, S.: Mathematical aspects of concept analysis. J. Math. Sci. **80**(2), 1654–1698 (1996)

19. Kuznetsov, S.O.: Fitting pattern structures to knowledge discovery in big data. In: Cellier, P., Distel, F., Ganter, B. (eds.) ICFCA 2013. LNCS, vol. 7880, pp. 254–266. Springer, Heidelberg (2013)

20. Kuznetsov, S.O., Poelmans, J.: Knowledge representation and processing with formal concept analysis. Wiley Interdisc. Rew: Data Min. Knowl. Discov. **3**(3), 200–215 (2013)

21. Lopes, S., Petit, J.-M., Lakhal, L.: Functional and approximate dependency mining: database and fca points of view. J. Exp. Theor. Artif. Intell. **14**(2–3), 93–114 (2002)

22. Medina, R., Nourine, L.: A unified hierarchy for functional dependencies, conditional functional dependencies and association rules. In: Ferré, S., Rudolph, S. (eds.) ICFCA 2009. LNCS, vol. 5548, pp. 98–113. Springer, Heidelberg (2009)

23. Nedjar, S., Pesci, F., Lakhal, L., Cicchetti, R.: The agree concept lattice for multidimensional database analysis. In: Jäschke, R. (ed.) ICFCA 2011. LNCS, vol. 6628, pp. 219–234. Springer, Heidelberg (2011)

24. Poelmans, J., Ignatov, D.I., Kuznetsov, S.O., Dedene, G.: Formal concept analysis in knowledge processing: a survey on applications. Expert Syst. Appl. **40**(16), 6538–6560 (2013)

25. Poelmans, J., Kuznetsov, S.O., Ignatov, D.I., Dedene, G.: Formal concept analysis in knowledge processing: aD survey on models and techniques. Expert Syst. Appl. **40**(16), 6601–6623 (2013)

26. Simovici, D., Jaroszewicz, S.: An axiomatization of partition entropy. IEEE Trans. Inf. Theory **48**(7), 2138–2142 (2002)

27. Simovici, D.A., Cristofor, D., Cristofor, L.: Impurity measures in databases. Acta Inf. **38**(5), 307–324 (2002)

28. Song, S., Chen, L.: Differential dependencies: reasoning and discovery. ACM Trans. Database Syst. **36**(3), 16:1–16:41 (2011)

29. Song, S., Chen, L.: Efficient discovery of similarity constraints for matching dependencies. Data Knowl. Eng. **87**, 146–166 (2013)

30. Song, S., Chen, L., Yu, P.S.: Comparable dependencies over heterogeneous data. The VLDB J. **22**(2), 253–274 (2013)

31. Ullman, J.: Principles of Database Systems and Knowledge-Based Systems, vol. 1–2. Computer Science Press, Rockville (MD) (1989)

32. Valtchev, P., Missaoui, R., Godin, R.: Formal concept analysis for knowledge discovery and data mining: the new challenges. In: Eklund, P. (ed.) ICFCA 2004. LNCS (LNAI), vol. 2961, pp. 352–371. Springer, Heidelberg (2004)

33. Wille, R.: Why can concept lattices support knowledge discovery in databases? J. Exp. Theor. Artif. Intell. **14**(2–3), 81–92 (2002)

34. Wyss, C.M., Giannella, C.M., Robertson, E.L.: FastFDs: a heuristic-driven, depth-first algorithm for mining functional dependencies from relation instances - extended abstract. In: Kambayashi, Y., Winiwarter, W., Arikawa, M. (eds.) DaWaK 2001. LNCS, vol. 2114, pp. 101–110. Springer, Heidelberg (2001)

35. Yao, H., Hamilton, H.J.: Mining functional dependencies from data. Data Min. Knowl. Discov. **16**(2), 197–219 (2008)

Review of Relation Extraction Methods: What Is New Out There?

Natalia Konstantinova[(✉)]

University of Wolverhampton, Wolverhampton, UK
n.konstantinova@wlv.ac.uk

Abstract. Relation extraction is a part of Information Extraction and an established task in Natural Language Processing. This paper presents an overview of the main directions of research and recent advances in the field. It reviews various techniques used for relation extraction including knowledge-based, supervised and self-supervised methods. We also mention applications of relation extraction and identify current trends in the way the field is developing.

Keywords: Relation extraction · Information extraction · Natural language processing · Review

1 Introduction

The modern world is rapidly developing and, in order to keep up-to-date, people must process large volume of information every day. Not only is the amount of this information is constantly increasing but the type of information is changing all the time. As a consequence of the sheer volume and heterogeneous nature of the information it is becoming impossible to analyse this data manually and new techniques are being used to automate this process. The field of Natural Language Processing (NLP) addresses this issue by analysing texts written in natural language and trying to understand them and extract valuable information. The problem of obtaining structured information from the text is dealt by Information Extraction (IE), a field of NLP. In this paper we mainly focus on one stage of IE – Relation Extraction (RE).

This paper is organised in the following way: Sect. 2 describes in more detail the field of Information Extraction and provides background on all its stages, Sect. 3 introduces the task of Relation Extraction, Sect. 4 presents knowledge-based methods, Sect. 5 describes supervised methods and Sect. 6 provides more details about the self-supervised approach. Section 7 introduces relation extraction as a part of joint modelling of several stages of IE. The paper finishes with Sect. 8 which summarises the material presented in the paper and also highlights possible future development of the field.

© Springer International Publishing Switzerland 2014
D.I. Ignatov et al. (Eds.): AIST 2014, CCIS 436, pp. 15–28, 2014.
DOI: 10.1007/978-3-319-12580-0_2

2   Information Extraction

Information extraction (IE) is a field of computational linguistics which plays a crucial role in the efficient management of data. It is defined as "a process of getting structured data from unstructured information in the text" (Jurafsky and Martin 2009). Grishman (1997) describes this process as "the identification of instances of a particular class of events or relationships in a natural language text, and the extraction of the relevant arguments of the event or relationship". After the information is structured and added to a database it can be used by a wide range of NLP applications, including information retrieval, question answering and many others.

Information extraction challenge has a long history and goes back to the late 1970s (Cowie and Lehnert 1996); however the first commercial systems appeared only in the 1990s, e.g. JASPER (Andersen et al. 1992), specially built for Reuters. Later research was greatly inspired by a series of Message Under-standing Conferences (MUC)[1], which were initiated and financed by the Defense Advanced Research Projects Agency (DARPA) to encourage the development of new methods in information extraction. The importance of the MUCs was not the conferences themselves, but the evaluations and evaluation competitions they proposed (Grishman and Sundheim 1996). The organisers of these confer-ences defined tasks for all the participants, prepared the data and developed the evaluation framework for each task. Researchers had to address the task and find the best solution; therefore it added competition element to the research. In addi-tion to all the above-mentioned advantages, these events were an opportunity to get comparable results and evaluate objectively the performance of different systems. MUCs were followed by several ACE (Automatic Content Extraction)[2] evaluations which also provided valuable feedback for researchers.

Usually IE, as many other NLP tasks, can be regarded as a pipeline process, where some kind of information is extracted at each stage. Jurafsky and Martin (2009) point out several different types of information that can be extracted:

- named entities (NE);
- temporal expressions;
- numeric values;
- relations between entities and expressions previously identified;
- events/template filling.

Generally IE starts with the detection and classification of proper names found in the text, which is usually referred to as Named Entity Recognition (NER). Most commonly IE systems search for names of people, companies and organisations, and geographical places. But the choice of the precise kind of NE to be extracted depends greatly on the task and system in mind. Sometimes the notion of Named Entities is extended to include items that are not really names or entities, but bear important information for analysing the texts; therefore,

---

[1] http://www.itl.nist.gov/iaui/894.02/related_projects/muc/
[2] http://www.itl.nist.gov/iad/894.01/tests/ace/

numeric values, such as measurements and prices, or temporal expressions can be included in this category. Extraction of such kinds of data is extremely important for correct analysis of texts and reasoning.

Usually the next step in IE is coreference resolution, the identification of identity relations between Named Entities (Jurafsky and Martin 2009). At this stage, mentions of the same Named Entity, which are expressed using different linguistic realisations, are found. The process of coreference resolution is crucial for getting more accurate results in IE and more details about this process are provided in the next section.

Relation extraction is a step further in analysing information in the texts and turning unstructured information into structured information. This stage involves identifying the links between Named Entities and deciding which ones are meaningful for the concrete application or problem.

The final stage of information extraction is template filling. Template filling involves extracting appropriate material to fill in the slots in templates for some stereotypical situations that recur quite often. For example, we can be interested in extracting information about some terrorist attack and this event can be treated as a template, which has predefined slots: place, date, number of people injured/killed, organisation who took responsibility for the terrorist act, etc.

3   Relation Extraction

As mentioned in Sect. 2, relation extraction (RE) is one of the steps of information extraction. It typically follows named entity recognition and coreference resolution and aims to gather relations between NEs. Culotta et al. (2006) define relation extraction as:

> "the task of discovering semantic connections between entities. In text, this usually amounts to examining pairs of entities in a document and determining (from local language cues) whether a relation exists between them."

Nowadays there are a lot of systems extracting relations from texts and there are different methods for dealing with this problem. Etzioni et al. (2008) classify all the methods used for relation extraction into three classes:

- knowledge-based methods;
- supervised methods;
- self-supervised methods.

Each of these classes are explained in the remainder of this paper.

4   Knowledge-Based Methods

The first category of methods is used usually in domain-specific tasks, where the texts are similar and a closed set of relations needs to be identified. Systems which use these methods rely on pattern-matching rules manually crafted for each domain (Riloff and Jones 1999; Pasca 2004). However, not all the relations are domain-dependent and there are some domain-independent ones. Hearst (1992) describes the usage of lexico-syntactic patterns for extraction of hyponymy relations in an open domain. These patterns capture such hyponymy relations as between *"author"* and *"Shakespeare"*, *"wound"* and *"injury"*, *"England"* and *"European country"*. However, the author notes that this method does not work well for some other kinds of relations, for example, meronymy. This is explained by the fact that patterns do not tend to uniquely identify the given relation.

The systems which participated in MUC and deal with relation extraction also rely on rich rules for identifying relations (Fukumoto et al. 1998; Garigliano et al. 1998; Humphreys et al. 1998). Humphreys et al. (1998) mention that they tried to add only those rules which were (almost) certain never to generate errors in analysis; therefore, they had adopted a low recall and high precision approach. However, in this case, many relations may be missed due to the lack of unambiguous rules to extract them.

To conclude, knowledge-based methods are not easily portable to other domains and involve too much manual labour. However, they can be used effectively if the main aim is to get results quickly in well-defined domains and document collections.

5   Supervised Methods

Supervised methods rely on a training set where domain-specific examples have been tagged. Such systems automatically learn extractors for relations by using machine-learning techniques. The main problem of using these methods is that the development of a suitably tagged corpus can take a lot of time and effort. On the other hand, these systems can be easily adapted to a different domain provided there is training data.

There are different ways that extractors can be learnt in order to solve the problem of supervised relation extraction: kernel methods (Zhao and Grishman 2005; Bunescu and Mooney 2006), logistic regression (Kambhatla 2004), augmented parsing (Miller et al. 2000), Conditional Random Fields (CRF) (Culotta et al. 2006).

In RE in general and supervised RE in particular a lot of research was done for IS-A relations and extraction of taxonomies. Several resources were built based on collaboratively built Wikipedia (YAGO – (Suchanek et al. 2007); DBpedia – (Auer et al. 2007); Freebase – (Bollacker et al. 2008); WikiNet – (Nastase et al. 2010)). In general, Wikipedia is becoming more and more popular as a source for RE, e.g. (Ponzetto and Strube 2007; Nguyen et al. 2007a, b, c). Query logs are also considered a valuable source of information for RE and their analysis is even argued to give better results than other suggested methods in the field (Paşca 2007, 2009).

5.1 Weakly-Supervised Methods

Some supervised systems also use bootstrapping to make construction of the training data easier. These methods are also sometimes referred to as "weakly-supervised information extraction". Brin (1998) describes the *DIPRE* (Dual Iterative Pattern Relation Expansion) method used for identifying authors of the books. It uses an initial small set of seeds or a set of hand-constructed extraction patterns to begin the training process. After the occurrences of needed information are found, they are further used for recognition of new patterns. Regardless of how promising bootstrapping can seem, error propagation becomes a serious problem: mistakes in extraction at the initial stages generate more mistakes at later stages and decrease the accuracy of the extraction process. For example, errors that expand to named entity recognition, e.g. extracting incomplete proper names, result in choosing incorrect seeds for the next step of bootstrapping. Another problem that can occur is that of semantic drift. This happens when senses of the words are not taken into account and therefore each iteration results in a move from the original meaning. Some researchers (Kozareva and Hovy 2010; Hovy et al. 2009; Kozareva et al. 2008) have suggested ways to avoid this problem and enhance the performance of this method by using doubly-anchored patterns (which include both the class name and a class member) as well as graph structures. Such patterns have two anchor seed positions "{type} such as {seed} and *" and also one open position for the terms to be learnt, for example, pattern "Presidents such as Ford and {X}" can be used to learn names of the presidents. Graphs are used for storing information about patterns, found words and links to entities they helped to find. This data is further used for calculating popularity and productivity of the candidate words. This approach helps to enhance the accuracy of bootstrapping and to find high-quality information using only a few seeds. Kozareva (2012) employs a similar approach for the extraction of cause-effect relations, where the pattern for bootstrapping has a form of "X and Y verb Z", for example, "* and virus cause *". Human-based evaluation reports 89 % accuracy on 1500 examples.

6 Self-supervised Systems

Self-supervised systems go further in making the process of information extraction unsupervised. The KnowItAll Web IE system (Etzioni et al. 2005), an example of a self-supervised system, learns "to label its own training examples using only a small set of domain-independent extraction patterns". It uses a set of generic patterns to automatically instantiate relation-specific extraction rules and then learns domain-specific extraction rules and the whole process is repeated iteratively.

The Intelligence in Wikipedia (IWP) project (Weld et al. 2008) is another example of a self-supervised system. It bootstraps from the Wikipedia corpus, exploiting the fact that each article corresponds to a primary object and that many articles contain infoboxes (brief tabular information about the article). This system is able to use Wikipedia infoboxes as a starting point for training the classifiers for the page type. IWP trains extractors for the various attributes and they can later be used for extracting information from general Web pages. The disadvantage of IWP is that the amount of relations described in Wikipedia infoboxes is limited and so not all relations can be extracted using this method.

6.1 Open Information Extraction

Etzioni et al. (2008) introduced the notion of Open Information Extraction, which is opposed to Traditional Relation Extraction. Open information extraction is "a novel extraction paradigm that tackles an unbounded number of relations". This method does not presuppose a predefined set of relations and is targeted at all relations that can be extracted.

The Open Relation extraction approach is relatively a new one, so there is only a small amount of projects using it. TextRunner (Banko and Etzioni 2008; Banko et al. 2007) is an example of such a system. A set of relation-independent lexico-syntactic patterns is used to build a relation-independent extraction model. It was found that 95 % of all relations in English can be described by only 8 general patterns, e.g. *"E1 Verb E2"*. The input of such a system is only a corpus and some relation-independent heuristics, relation names are not known in advance. Conditional Random Fields (CRF) are used to identify spans of tokens believed to indicate explicit mentions of relationships between entities and the whole problem of relation extraction is treated as a problem of sequence labelling. The set of linguistic features used in this system is similar to those used by other state of-the-art relation extraction systems and includes e.g. part-of-speech tags, regular expressions for detection of capitalization and punctuation, context words. At this stage of development this system "is able to extract instances of the four most frequently observed relation types: Verb, Noun+Prep, Verb+Prep and Infinitive". It has a number of limitations, which are however common to all RE systems: it extracts only explicitly expressed relations that are primarily word-based; relations should occur between entity names within the same sentence.

Banko and Etzioni (2008) report a precision of 88.3 % and a recall of 45.2 %. Even though the system shows very good results the relations are not specified and so there are difficulties in using them in some other systems. Output of the system consists of tuples stating there is some relation between two entities, but there is no generalization of these relations.

Wu and Weld (2010) combine the idea of Open Relation Extraction and the use of Wikipedia infoboxes and produce systems called WOE^{parse} and WOE^{pos}. WOE^{parse} improves TextRunner dramatically but it is 30 times slower than TextRunner. However, WOE^{pos} does not have this disadvantage and still shows an improved F-measure over TextRunner between 15 % to 34 % on three corpora.

Fader et al. (2011) identify several flaws in previous works in Open Information Extraction: "the learned extractors ignore both "holistic" aspects of the relation phrase (e.g., is it contiguous?) as well as lexical aspects (e.g., how many instances of this relation are there?)". They target these problems by introducing syntactic constraints (e.g., they require the relation phrase to match the POS tag pattern) and lexical constraints. Their system ReVerb achieves an AUC which is 30 % better than WOE (Wu and Weld 2010) and TextRunner (Banko and Etzioni 2008).

Nakashole et al. (2012a) approach this problem from another angle. They try to mine for patterns expressing various relations and organise then in hierarchies. They explore binary relations between entities and employ frequent itemset mining (Agrawal et al. 1993; Srikant and Agrawal 1996) to identify the most frequent patterns. Their work results in a resource called PATTY which contains 350.569 pattern synsets and subsumption relations and achieves 84.7 % accuracy. Unlike ReVerb (Fader et al. 2011) which constrains patterns to verbs or verb phrases that end with prepositions, PATTY can learn arbitrary patterns. The authors employ so called syntactic-ontologic-lexical patterns (SOL patterns). These patterns constitute a sequence of words, POS-tags, wildcards, and ontological types. For example, the pattern "persons [adj] voice * song" would match the strings my Winehouses soft voice in Rehab and Elvis Presleys solid voice in his song All shook up. Their approach is based on collecting dependency paths from the sentences where two named entities are tagged (YAGO2 (Hoffart et al. 2011) is used as a database of all NEs). Then the textual pattern is extracted by finding the shortest paths connecting two entities. All of these patterns are transformed into SOL (abstraction of a textual pattern). Frequent itemset technique is used for this: all textual patterns are decomposed into n-grams (n consecutive words). A SOL pattern contains only the n-grams that appear frequently in the corpus and the remaining word sequences are replaced by wildcards. The support set of the pattern is described as the set of pairs of entities that appear in the place of the entity placeholders in all strings in the corpus that match the pattern. The patterns are connected in one synset (so are considered synonymous) if their supporting sets coincide. The overlap of the supporting sets is also employed to identify subsumption relations between various synsets.

6.2 Distant Learning

Mintz et al. (2009) introduce a new term "distant supervision". The authors use a large semantic database Freebase containing 7,300 relations between 9 million named entities. For each pair of entities that appears in Freebase relation, they identify all sentences containing those entities in a large unlabeled corpus. At the next step textual features to train a relation classifier are extracted. Even though the 67,6 % of precision achieved using this method has room for improvement, it has inspired many researchers to further investigate in this direction.

Currently there are a number of papers trying to enhance "distant learning" in several directions. Some researchers target the heuristics that are used to map the relations in the databases to the texts, for example, (Takamatsu et al. 2012) argue that improving matching helps to make data less noisy and therefore enhances the quality of relation extraction in general.

Yao et al. (2010) propose using an undirected graphical model for relation extraction which employs "distant learning" but enforces selectional preferences. Riedel et al. (2010) reports 31 % error reduction compared to (Mintz et al. 2009).

Another problem that has been addressed is language ambiguity (Yao et al. 2011, 2012). Most methods cluster shallow or syntactic patterns of relation mentions, but consider only one possible sense per pattern. However, this assumption is often violated in reality. Yao et al. (2011) uses generative probabilistic models, where both entity type constraints within a relation and features on the dependency path between entity mentions are exploited. This research is similar to DIRT (Lin and Pantel 2001) which explores distributional similarity of dependency paths in order to discover different representations of the same semantic relation. However, Yao et al. (2011) employ another approach and apply LDA (Blei et al. 2003) with a slight modification: observations are relation tuples and not words. So as a result of this modification instead of representing semantically related words, the topic latent variable represents a relation type. The authors combine three models: Rel-LDA, Rel-LDA1 and Type-LDA. In the third model the authors split the features of a tuple into relation level features and entity level features. Relation level features include the dependency path, trigger, lexical and POS features; entity level features include the entity mention itself and its named entity tag. These models output clustering of observed relation tuples and their associated textual expressions. The evaluation shows that the use of these resulting clusters helps to improve distant learning and results in 12 % better performance.

Distant learning and other types of relation extraction are, as we have already seen, based on several assumptions. Another assumption that is often used in this field is that a pair of entities can have only one relation. However, if we examine the following examples – "Steve Jobs founded Apple" and "Steve Jobs is CEO of Apple" – we can see that this assumption is rather restrictive.

Hoffmann et al. (2011) identified this problem with previous RE systems and try to address this issue by employing Multi-Instance Multi-label (MIML) approach. They employ distant learning with Multi-Instance learning with overlapping relations (where two same instances may be in two different relations). The resulting system MultiR achieves competitive or higher precision over all ranges of recall.

Surdeanu et al. (2012) tackle the same problem. They identify two main problems of distant learning: (1) some training examples obtained through this heuristic are not valid (they report 31 %), (2) the same pair of entities can have several relations. Therefore they try to improve distant learning by taking into account Multi-instance Multi-label settings and using Bayesian framework (they call their system MIML-RE) which can capture dependencies between labels and learn in the presence of incorrect and incomplete labels.

When using "distant supervision" in its original version we are limited by the schema imposed by the database that is used for mapping. Yao et al. (2013) suggest several ways how it can be overcome. They suggest using raw texts in addition to distant supervision, therefore relations in the text and pre-existing structured databases can be employed together. Riedel et al. (2013) address this problem by using matrix factorisation and collaborative filtering. Previously, matrix factorisation was employed by Nickel et al. (2012) in order to predict new relations (triples) in terms of YAGO2. All relations between entities are presented as a matrix where there is an indication whether there is a relation or not. Riedel et al. (2013) use three models: (1) latent feature model, which is a generalised PCA (Collins et al. 2001); (2) neighbourhood model, neighbour based approach (Koren 2008); (3) entity model, which learns a latent entity representation from data. The authors also present a combined model that incorporates all three models with various weights. In order to overcome the lack of negative examples, they employ the technique of implicit feedback (Rendle et al. 2009), where observed true facts are given higher scores than unobserved (true or false) facts. The authors report competitive results of evaluation and also mention computational efficiency of their methods which is an important aspect for such systems. They also discuss the fact that this approach is not merely a tool for information extraction and that the same technique can be used for integrating databases with different schemata.

7   Joint Prediction

The joint modelling of several levels of Information extraction is also explored by several research papers. In their position paper Mccallum and Jensen (2003) propose to use "unified, relational, undirected graphical models for information extraction and data mining". This common inference procedure can help to improve all the stages, so that each component is able to make up for the weaknesses of the other and therefore improve the performance of both.

Finkel et al. (2006) explore the idea of joint modeling as well. They present a novel architecture, which models pipelines as Bayesian networks. Each low level task corresponds to a variable in the network, and then an approximate inference is performed to find the best labelling. This approach is tested on two tasks: semantic role labelling and recognizing textual entailment.

Roth and Yih (2007) employ the same idea when they combine two stages on Information Extraction: named entity recognition and relation extraction. However, Singh et al. (2013) go even further and include coreference resolution as well. So they propose a single, joint graphical model that represents the various dependencies between the tasks (entity tagging, relation extraction, and coreference). Their joint modelling approach helps to avoid cascading errors. The joint model obtains 12 % error reduction on tagging over the isolated models.

8   Conclusions

This paper introduced a field of Information Extraction and provided more details about recent developments in its subfield, Relation Extraction. We have presented the main approaches to this task and also outlined some challenges. All the methods described above have advantages and disadvantages and the choice depends greatly on the task in mind and the accuracy needed. Relation extraction has a lot of uses in NLP and can be beneficial for: semantic search, machine reading, question answering, knowledge harvesting, paraphrasing, building thesauri etc. (Nakashole et al. 2012b, 2013).

The field appears to be becoming more and more interdisciplinary and methods from data mining and recommendation systems are currently used to assist in the task of relation extraction (Cergani and Miettinen 2013; Riedel et al. 2013; Nakashole et al. 2012a). Also modeling all stages of Information Extraction as a single task is another recent trend in the field (Mccallum and Jensen 2003; Finkel et al. 2006; Roth and Yih 2007; Singh et al. 2013). The research in this area reports significant improvement in all tasks when they are modeled jointly, it also helps to avoid error propagation which is a frequent problem in the pipeline approach.

Research in terms of relation extraction has still room for improvement, however, it targets a very difficult problem where language ambiguity is a significant obstacle. The majority of research in the field is done for English language, therefore targeting other languages and exploring further multilingual information extraction and possibility of aligning resources in various languages can be the future direction of Relation Extraction.

References

Agrawal, R., Imieliński, T., Swami, A.: Mining association rules between sets of items in large databases. In: Proceedings of the 1993 ACM SIGMOD International Conference on Management of Data, SIGMOD '93, pp. 207–216. ACM, New York (1993)

Andersen, P.M., Hayes, P.J., Huettner, A.K., Schmandt, L.M., Nirenburg, I.B., Weinstein, S.P.: Automatic extraction of facts from press releases to generate news stories. In: Proceedings of the Third Conference on Applied Natural Language Processing, pp. 170–177 (1992)

Auer, S., Bizer, C., Kobilarov, G., Lehmann, J., Cyganiak, R., Ives, Z.G.: DBpedia: a nucleus for a web of open data. In: Aberer, K., Choi, K.-S., Noy, N., Allemang, D., Lee, K.-I., Nixon, L.J.B., Golbeck, J., Mika, P., Maynard, D., Mizoguchi, R., Schreiber, G., Cudré-Mauroux, P. (eds.) ASWC 2007 and ISWC 2007. LNCS, vol. 4825, pp. 722–735. Springer, Heidelberg (2007)

Banko, M., Etzioni, O.: The tradeoffs between open and traditional relation extraction. In: Proceedings of ACL-08: HLT, pp. 28–36 (2008)

Banko, M., Cafarella, M.J., Soderland, S., Broadhead, M., Etzioni, O.: Open information extraction from the web. In: Veloso, M.M. (ed.) Proceedings of the 20th International Joint Conference on Artificial Intelligence, Hyderabad, India, pp. 2670–2676 (2007)

Blei, D.M., Ng, A.Y., Jordan, M.I.: Latent dirichlet allocation. J. Mach. Learn. Res. 3, 993–1022 (2003)

Bollacker, K., Evans, C., Paritosh, P., Sturge, T., Taylor, J.: Freebase: a collaboratively created graph database for structuring human knowledge. In: Proceedings of the 2008 ACM SIGMOD International Conference on Management of Data, SIGMOD '08, pp. 1247–1250. ACM, New York (2008)

Brin, S.: Extracting patterns and relations from the world wide web. In: Proceedings of the First International Workshop on the Web and Databases, pp. 172–183 (1998)

Bunescu, R., Mooney, R.: Subsequence kernels for relation extraction. In: Weiss, Y., Schölkopf, B., Platt, J. (eds.) Advances in Neural Information Processing Systems 18, pp. 171–178. MIT Press, Cambridge (2006)

Cergani, E., Miettinen, P.: Discovering relations using matrix factorization methods. In: He, Q., Iyengar, A., Nejdl, W., Pei, J., Rastogi, R. (eds.) CIKM, pp. 1549–1552. ACM (2013)

Collins, M., Dasgupta, S., Schapire, R.E.: A generalization of principal component analysis to the exponential family. In: Leen, T., Dietterich, T., Tresp, V. (eds.) Advances in Neural Information Processing Systems. MIT Press, Cambridge (2001)

Cowie, J., Lehnert, W.: Information extraction. Commun. ACM **39**(1), 80–91 (1996)

Culotta, A., McCallum, A., Betz, J.: Integrating probabilistic extraction models and data mining to discover relations and patterns in text. In: Proceedings of the Main Conference on Human Language Technology Conference of the North American Chapter of the Association of Computational Linguistics, New York, pp. 296–303. Association for Computational Linguistics (2006)

Etzioni, O., Cafarella, M., Downey, D., Popescu, A.M., Shaked, T., Soderland, S., Weld, D.S., Yates, A.: Unsupervised named-entity extraction from the web: an experimental study. Artif. Intell. **165**, 91–134 (2005). (Elsevier Science Publishers Ltd., Essex, UK)

Etzioni, O., Banko, M., Soderland, S., Weld, D.S.: Open information extraction from the web. Commun. ACM **51**, 68–74 (2008)

Fader, A., Soderland, S., Etzioni, O.: Identifying relations for open information extraction. In: Proceedings of EMNLP 2011, UK, Edinburgh (2011)

Finkel, J.R., Manning, C.D., Ng, A.Y.: Solving the problem of cascading errors: approximate bayesian inference for linguistic annotation pipelines. In: Proceedings of the 2006 Conference on Empirical Methods in Natural Language Processing, EMNLP '06, pp. 618–626. Association for Computational Linguistics, Stroudsburg (2006)

Fukumoto, J., Masui, F., Shimohata, M., Sasaki, M.: Oki electric industry: description of the Oki system as used for MUC-7. In: Proceedings of the 7th Message Understanding Conference (1998)

Garigliano, R., Urbanowicz, A., Nettleton, D.J.: University of Durham: description of the LOLITA system as used in MUC-7. In: Proceedings of the 7th Message Understanding Conference (1998)

Grishman, R.: Information extraction: techniques and challenges. In: Pazienza, M.T. (ed.) SCIE 1997. LNCS, vol. 1299, pp. 10–27. Springer, Heidelberg (1997)

Grishman, R., Sundheim, B.: Message understanding conference-6: a brief history. In: Proceedings of the 16th Conference on Computational Linguistics, pp. 466–471. Association for Computational Linguistics, Morristown (1996)

Hearst, M.A.: Automatic acquisition of hyponyms from large text corpora. In: Proceedings of the 14th Conference on Computational Linguistics, pp. 539–545. Association for Computational Linguistics, Morristown (1992)

Hoffart, J., Suchanek, F.M., Berberich, K., Lewis-Kelham, E., de Melo, G., Weikum, G.: Yago2: exploring and querying world knowledge in time, space, context, and many languages. In: Proceedings of the 20th International Conference Companion on World Wide Web, WWW '11, pp. 229–232. ACM, New York (2011)

Hoffmann, R., Zhang, C., Ling, X., Zettlemoyer, L., Weld, D.S.: Knowledge-based weak supervision for information extraction of overlapping relations. In: Proceedings of the 49th Annual Meeting of the Association for Computational Linguistics: Human Language Technologies - Volume 1, HLT '11, pp. 541–550. Association for Computational Linguistics, Stroudsburg (2011)

Hovy, E., Kozareva, Z., Riloff, E.: Toward completeness in concept extraction and classification. In: EMNLP '09: Proceedings of the 2009 Conference on Empirical Methods in Natural Language Processing, pp. 948–957. Association for Computational Linguistics, Morristown (2009)

Humphreys, K., Gaizauskas, R., Azzam, S., Huyck, C., Mitchell, B., Cunningham, H., Wilks, Y.: University of Sheffield: description of the LaSIE-II system as used for MUC-7. In: Proceedings of MUC-7 (1998)

Jurafsky, D., Martin, J.H.: Speech and language processing an introduction to natural language processing, computational linguistics, and speech recognition, 2nd edn. Prentice-Hall Inc., Upper Saddle River (2009)

Kambhatla, N.: Combining lexical, syntactic, and semantic features with maximum entropy models for extracting relations. In: Proceedings of the ACL 2004 on Interactive poster and demonstration sessions, p. 22. Association for Computational Linguistics, Morristown (2004)

Koren, Y.: Factorization meets the neighborhood: a multifaceted collaborative filtering model. In: Proceedings of the 14th ACM SIGKDD International Conference on Knowledge Discovery and Data Mining, KDD '08, pp. 426–434. ACM, New York (2008)

Kozareva, Z.: Cause-effect relation learning. In: Workshop Proceedings of TextGraphs-7 on Graph-based Methods for Natural Language Processing, TextGraphs-7 '12, pp. 39–43. Association for Computational Linguistics, Stroudsburg (2012)

Kozareva, Z., Hovy, E.: Learning arguments and supertypes of semantic relations using recursive patterns. In: Proceedings of the 48th Annual Meeting of the Association for Computational Linguistics, Uppsala, Sweden, pp. 1482–1491. Association for Computational Linguistics (2010)

Kozareva, Z., Riloff, E., Hovy, E.: Semantic class learning from the web with hyponym pattern linkage graphs. In: Proceedings of ACL-08: HLT, Columbus, Ohio, pp. 1048–1056. Association for Computational Linguistics (2008)

Lin, D., Pantel, P.: DIRT - discovery of inference rules from text. In: Proceedings of the Seventh ACM SIGKDD International Conference on Knowledge Discovery and Data Mining, KDD '01, pp. 323–328. ACM, New York (2001)

Mccallum, A., Jensen, D.: A note on the unification of information extraction and data mining using conditional-probability, relational models. In: Proceedings of the IJCAI-2003 Workshop on Learning Statistical Models from Relational Data (2003)

Miller, S., Fox, H., Ramshaw, L., Weischedel, R.: A novel use of statistical parsing to extract information from text. In: Proceedings of the 1st North American Chapter of the Association for Computational Linguistics Conference, Seattle, Washington, pp. 226–233. Morgan Kaufmann Publishers Inc. (2000)

Mintz, M., Bills, S., Snow, R., Jurafsky, D.: Distant supervision for relation extraction without labeled data. In: Proceedings of the Joint Conference of the 47th Annual Meeting of the ACL and the 4th International Joint Conference on Natural Language Processing of the AFNLP: Volume 2 - Volume 2, ACL '09, pp. 1003–1011. Association for Computational Linguistics, Stroudsburg (2009)

Nakashole, N., Weikum, G., Suchanek, F.: Patty: a taxonomy of relational patterns with semantic types. In: Proceedings of the 2012 Joint Conference on Empirical Methods in Natural Language Processing and Computational Natural Language Learning, EMNLP-CoNLL '12, pp. 1135–1145. Association for Computational Linguistics, Stroudsburg (2012a)

Nakashole, N., Weikum, G., Suchanek, F.M.: Discovering and exploring relations on the web. PVLDB 5(12), 1982–1985 (2012b)

Nakashole, N., Weikum, G., Suchanek, F.M.: Discovering semantic relations from the web and organizing them with patty. SIGMOD Rec. **42**(2), 29–34 (2013)

Nastase, V., Strube, M., Boerschinger, B., Zirn, C., Elghafari, A.: Wikinet: a very large scale multi-lingual concept network. In: Calzolari, N., Choukri, K., Maegaard, B., Mariani, J., Odijk, J., Piperidis, S., Rosner, M., Tapias, D. (eds.) LREC. European Language Resources Association (2010)

Nguyen, D.P., Matsuo, Y., Ishizuka, M.: Exploiting syntactic and semantic information for relation extraction from Wikipedia. In: Proceedings of the IJCAI Workshop on Text-Mining and Link- Analysis, TextLink07 (2007a)

Nguyen, D.P., Matsuo, Y., Ishizuka, M.: Relation extraction from Wikipedia using subtree mining. In: Proceedings of the Twenty-Second AAAI Conference on Artificial Intelligence, Vancouver, British Columbia, Canada, pp. 1414–1420. AAAI Press (2007b)

Nguyen, D.P.T., Matsuo, Y., Ishizuka, M.: Subtree mining for relation extraction from Wikipedia. In: Sidner, C.L., Schultz, T., Stone, M., Zhai, C. (eds.) Human Language Technology Conference of the North American Chapter of the Association of Computational Linguistics, Proceedings, Rochester, New York, USA, pp. 125–128. The Association for Computational Linguistics (2007c)

Nickel, M., Tresp, V., Kriegel, H.P.: Factorizing YAGO: scalable machine learning for linked data. In: Proceedings of the 21st International Conference on World Wide Web, WWW '12, pp. 271–280. ACM, New York (2012)

Paşca, M.: Organizing and searching the world wide web of facts - step two: harnessing the wisdom of the crowds. In: WWW '07: Proceedings of the 16th International Conference on World Wide Web, pp. 101–110. ACM, New York (2007)

Paşca, M.: Outclassing Wikipedia in open-domain information extraction: weakly-supervised acquisition of attributes over conceptual hierarchies. In: Proceedings of the 12th Conference of the European Chapter of the Association for Computational Linguistics, Athens, Greece, pp. 639–647. Association for Computational Linguistics (2009)

Pasca, M.: Acquisition of categorized named entities for web search. In: Proceedings of the Thirteenth ACM International Conference on Information and Knowledge Management, CIKM '04, pp. 137–145. ACM, New York (2004)

Ponzetto, S.P., Strube, M.: Deriving a large scale taxonomy from Wikipedia. In: Proceedings of the 22nd Conference on the Advancement of Artificial Intelligence, Vancouver, B.C., Canada, pp. 1440–1445 (2007)

Rendle, S., Freudenthaler, C., Gantner, Z., Schmidt-Thieme, L.: BPR: Bayesian personalized ranking from implicit feedback. In: Proceedings of the Twenty-Fifth Conference on Uncertainty in Artificial Intelligence, UAI '09, Arlington, Virginia, United States, pp. 452–461. AUAI Press (2009)

Riedel, S., Yao, L., McCallum, A.: Modeling relations and their mentions without labeled text. In: Balcázar, J.L., Bonchi, F., Gionis, A., Sebag, M. (eds.) ECML PKDD 2010, Part III. LNCS, vol. 6323, pp. 148–163. Springer, Heidelberg (2010)

Riedel, S., Yao, L., McCallum, A., Marlin, B.M.: Relation extraction with matrix factorization and universal schemas. In: Vanderwende, L., III, H.D., Kirchhoff, K. (eds.) HLT-NAACL, pp. 74–84. The Association for Computational Linguistics (2013)

Riloff, E., Jones, R: Learning dictionaries for information extraction by multi-level bootstrapping. In: Proceedings of the Sixteenth National Conference on Artificial Intelligence and the Eleventh Innovative Applications of Artificial Intelligence Conference, Menlo Park, CA, USA, AAAI '99/IAAI '99, pp. 474–479. American Association for Artificial Intelligence (1999)

Roth, D., Yih, W.: Global inference for entity and relation identification via a linear programming formulation. In: Getoor, L., Taskar, B. (eds.) Introduction to Statistical Relational Learning. MIT Press, Cambridge (2007)

Singh, S., Riedel, S., Martin, B., Zheng, J., McCallum, A.: Joint inference of entities, relations, and coreference. In: Proceedings of the 3rd Workshop on Automated Knowledge Base Construction (2013)

Srikant, R., Agrawal, R.: Mining sequential patterns: generalizations and performance improvements. In: Apers, P.M.G., Bouzeghoub, M., Gardarin, G. (eds.) EDBT 1996. LNCS, vol. 1057, pp. 3–17. Springer, Heidelberg (1996)

Suchanek, F.M., Kasneci, G., Weikum, G.: YAGO: a core of semantic knowledge. In: Proceedings of WWW-07, pp. 697–706 (2007)

Surdeanu, M., Tibshirani, J., Nallapati, R., Manning, C.D.: Multi-instance multi-label learning for relation extraction. In: Proceedings of the 2012 Joint Conference on Empirical Methods in Natural Language Processing and Computational Natural Language Learning, EMNLP-CoNLL '12, pp. 455–465. Association for Computational Linguistics, Stroudsburg (2012)

Takamatsu, S., Sato, I., Nakagawa, H.: Reducing wrong labels in distant supervision for relation extraction. In: Proceedings of the 50th Annual Meeting of the Association for Computational Linguistics: Long Papers - Volume 1, ACL '12, pp. 721–729. Association for Computational Linguistics, Stroudsburg (2012)

Weld, D.S., Wu, F., Adar, E., Amershi, S., Fogarty, J., Hoffmann, R., Patel, K., Skinner, M.: Intelligence in Wikipedia. In: Proceedings of the 23rd AAAI Conference, Chicago, USA (2008)

Wu, F., Weld, D.S.: Open information extraction using Wikipedia. In: Proceedings of the 48th Annual Meeting of the Association for Computational Linguistics, ACL '10, pp. 118–127. Association for Computational Linguistics, Stroudsburg (2010)

Yao, L., Riedel, S., McCallum, A.: Collective cross-document relation extraction without labelled data. In: EMNLP, pp. 1013–1023. ACL (2010)

Yao, L., Haghighi, A., Riedel, S., McCallum, A.: Structured relation discovery using generative models. In: Proceedings of the Conference on Empirical Methods in Natural Language Processing (EMNLP '11), pp. 1456–1466 (2011)

Yao, L., Riedel, S., McCallum, A.: Unsupervised relation discovery with sense disambiguation. In: ACL, The Association for Computer Linguistics, pp. 712–720 (2012)

Yao, L., Riedel, S., McCallum, A.: Universal schema for entity type prediction. In: Proceedings of the 3rd Workshop on Automated Knowledge Base Construction (2013)

Zhao, S., Grishman, R.: Extracting relations with integrated information using kernel methods. In: ACL '05: Proceedings of the 43rd Annual Meeting on Association for Computational Linguistics, pp. 419–426. Association for Computational Linguistics, Morristown (2005)

Tutorial on Probabilistic Topic Modeling: Additive Regularization for Stochastic Matrix Factorization

Konstantin Vorontsov[1](✉) and Anna Potapenko[2]

[1] The Higher School of Economics, Dorodnicyn Computing Centre of RAS, Moscow Institute of Physics and Technology, Moscow, Russia
voron@forecsys.ru
[2] Dorodnicyn Computing Centre of RAS, Moscow State University, Moscow, Russia
anya_potapenko@mail.ru

Abstract. Probabilistic topic modeling of text collections is a powerful tool for statistical text analysis. In this tutorial we introduce a novel non-Bayesian approach, called *Additive Regularization of Topic Models*. ARTM is free of redundant probabilistic assumptions and provides a simple inference for many combined and multi-objective topic models.

Keywords: Probabilistic topic modeling · Regularization of ill-posed inverse problems · Stochastic matrix factorization · Probabilistic latent sematic analysis · Latent Dirichlet Allocation · EM-algorithm

1 Introduction

Topic modeling is a rapidly developing branch of statistical text analysis [1]. Topic model uncovers a hidden thematic structure of the text collection and finds a highly compressed representation of each document by a set of its topics. From the statistical point of view, each topic is a set of words or phrases that frequently co-occur in many documents. The topical representation of a document captures the most important information about its semantics and therefore is useful for many applications including information retrieval, classification, categorization, summarization and segmentation of texts.

Hundreds of specialized topic models have been developed recently to meet various requirements coming from applications. For example, some of the models are capable to discover how topics evolve through time, how they are connected to each other, how they form topic hierarchies. Other models take into account additional information such as authors, sources, categories, citations or links between documents, or other kinds of document labels [2]. They can also be used to reveal the semantics of non-textual objects connected to the documents such as images, named entities or document users. Some of the models are focused on making topics more stable, sparse, robust, and better interpretable by humans. Linguistically motivated models benefit from syntactic considerations, grouping

© Springer International Publishing Switzerland 2014
D.I. Ignatov et al. (Eds.): AIST 2014, CCIS 436, pp. 29–46, 2014.
DOI: 10.1007/978-3-319-12580-0_3

words into n-grams, finding collocations or constituent phrases. More ideas and applications of topic modeling can be found in the survey [3].

A *probabilistic topic model* defines each topic by a multinomial distribution over words, and then describes each document with a multinomial distribution over topics. Most recent models are based on a mainstream topic model LDA, Latent Dirichlet Allocation [4]. LDA is a two-level Bayesian generative model, which assumes that topic distributions over words and document distributions over topics are generated from prior Dirichlet distributions. This assumption facilitates Bayesian inference due to the fact that the Dirichlet distribution is a conjugate to the multinomial one. However, the Dirichlet distribution has no convincing linguistic motivations and conflicts with two natural assumptions of sparsity: (1) most of the topics have zero probability in a document, and (2) most of the words have zero probability in a topic. The attempts to provide sparsity preserving Dirichlet prior lead to overcomplicated models [5–9]. Finally, Bayesian inference complicates the combination of many requirements into a single multi-objective topic model. The evolutionary algorithms recently proposed in [10] seem to be computationally infeasible for large text collections.

In this tutorial we present a survey of popular topic models in terms of a novel non-Bayesian approach — *Additive Regularization of Topic Models* (ARTM) [11], which removes the above limitations, simplifies theory without loss of generality, and reduces barriers to entry into topic modeling research field.

The motivations and essentials of ARTM may be briefly stated as follows. Learning of a topic model from a text collection is an ill-posed inverse problem of stochastic matrix factorization. Generally it has an infinite set of solutions. To choose a better solution we add a weighted sum of problem-oriented regularization penalty terms to the log-likelihood. Then the model inference in ARTM can be performed by a simple differentiation of the regularizers over model parameters. We show that many models, which previously required a complicated inference, can be obtained "in one line" within ARTM. The weights in a linear combination of regularizers can be adopted during the iterative process. Our experiments demonstrate that ARTM can combine regularizers that improve many criteria at once almost without a loss of the likelihood.

2   Topic Models PLSA and LDA

In this section we describe Probabilistic Latent Sematic Analysis (PLSA) model, which was historically a predecessor of LDA. PLSA is a more convenient starting point for ARTM because it does not have regularizers at all. We provide the Expectation-Maximization (EM) algorithm with an elementary explanation, then describe an experiment on the model data that shows the instability of both PLSA and LDA models. The non-uniqueness and the instability of the solution does motivate a problem-oriented additive regularization.

Model assumptions. Let D denote a set (collection) of texts and W denote a set (vocabulary) of all words from these texts. Note that vocabulary may contain keyphrases as well, but we will not distinguish them from single words.

Each document $d \in D$ is a sequence of n_d words $(w_1, \ldots, w_{n_d})$ from the vocabulary W. Each word might appear multiple times in the same document.

Assume that each word occurrence in each document refers to some latent topic from a finite set of topics T. Text collection is considered to be a sample of triples (w_i, d_i, t_i), $i = 1, \ldots, n$ drawn independently from a discrete distribution $p(w, d, t)$ over a finite probability space $W \times D \times T$. Words w and documents d are observable variables, while topics t are *latent* (hidden) variables.

Following the "bag of words" model, we represent each document by a subset of words $d \subset W$ and the corresponding integers n_{dw}, which count how many times the word w appears in the document d.

Conditional independence is an assumption that each topic generates words regardless of the document: $p(w \,|\, t) = p(w \,|\, d, t)$. According to the law of total probability and the assumption of conditional independence

$$p(w \,|\, d) = \sum_{t \in T} p(t \,|\, d) p(w \,|\, t). \tag{1}$$

The probabilistic model (1) describes how the collection D is generated from the known distributions $p(t \,|\, d)$ and $p(w \,|\, t)$. Learning a topic model is an inverse problem: to find distributions $p(t \,|\, d)$ and $p(w \,|\, t)$ given a collection D.

Stochastic matrix factorization. Our problem is equivalent to finding an approximate representation of observable data matrix

$$F = \left(f_{wd} \right)_{W \times D}, \quad f_{wd} = \hat{p}(w \,|\, d) = n_{dw}/n_d,$$

as a product $F \approx \Phi\Theta$ of two unknown matrices — the matrix Φ of *word probabilities for the topics* and the matrix Θ of *topic probabilities for the documents*:

$$\Phi = (\phi_{wt})_{W \times T}, \ \phi_{wt} = p(w \,|\, t), \ \phi_t = (\phi_{wt})_{w \in W};$$
$$\Theta = (\theta_{td})_{T \times D}, \quad \theta_{td} = p(t \,|\, d), \ \theta_d = (\theta_{td})_{t \in T}.$$

Matrices F, Φ and Θ are *stochastic*, that is, their columns f_d, ϕ_t, θ_d are non-negative and normalized representing discrete distributions. Usually the number of topics $|T|$ is much smaller than both $|D|$ and $|W|$.

Likelihood maximization. In probabilistic latent semantic analysis (PLSA) [12] the topic model (1) is learned by the log-likelihood maximization:

$$\ln \prod_{i=1}^{n} p(d_i, w_i) = \sum_{d \in D} \sum_{w \in d} n_{dw} \ln p(w \,|\, d) + \sum_{d \in D} n_d \ln p(d) \to \max,$$

which results in a constrained maximization problem:

$$L(\Phi, \Theta) = \sum_{d \in D} \sum_{w \in d} n_{dw} \ln \sum_{t \in T} \phi_{wt} \theta_{td} \to \max_{\Phi, \Theta}; \tag{2}$$

$$\sum_{w \in W} \phi_{wt} = 1, \quad \phi_{wt} \geq 0; \qquad \sum_{t \in T} \theta_{td} = 1, \quad \theta_{td} \geq 0. \tag{3}$$

Algorithm 2.1. The rational EM-algorithm for PLSA.

Input: document collection D, number of topics $|T|$, initialized Φ, Θ;
Output: Φ, Θ;

1 **repeat**
2   zeroize n_{wt}, n_{dt}, n_t, n_d for all $d \in D, w \in W, t \in T$;
3   **for all** $d \in D, w \in d$
4          $Z := \sum_{t \in T} \phi_{wt}\theta_{td}$;
5   **for all** $t \in T$: $\phi_{wt}\theta_{td} > 0$
6   increase n_{wt}, n_{dt}, n_t, n_d by $\delta = n_{dw}\phi_{wt}\theta_{td}/Z$;

7      $\phi_{wt} := n_{wt}/n_t$ for all $w \in W, t \in T$;
8      $\theta_{td} := n_{dt}/n_d$ for all $d \in D, t \in T$;
9 **until** Φ and Θ converge;

EM-algorithm. The problem (2), (3) can be solved by an iterative EM-algorithm. First, the columns of the matrices Φ and Θ are initialized with random distributions. Then two steps (E-step and M-step) are repeated in a loop.

At the E-step the probability distributions for the latent topics $p(t \mid d, w)$ are estimated for each word w in each document d using the Bayes' rule. Auxiliary variables n_{dwt} are introduced to estimate how many times the word w appears in the document d with relation to the topic t:

$$n_{dwt} = n_{dw}p(t \mid d, w), \quad p(t \mid d, w) = \frac{\phi_{wt}\theta_{td}}{\sum_{s \in T} \phi_{ws}\theta_{sd}}. \tag{4}$$

At the M-step summation of n_{dwt} values over d, w, t provides empirical estimates for the unknown conditional probabilities:

$$\phi_{wt} = \frac{n_{wt}}{n_t}, \qquad n_{wt} = \sum_{d \in D} n_{dwt}, \qquad n_t = \sum_{w \in W} n_{wt},$$

$$\theta_{td} = \frac{n_{dt}}{n_d}, \qquad n_{dt} = \sum_{w \in d} n_{dwt}, \qquad n_d = \sum_{t \in T} n_{dt},$$

which can be rewritten in a shorter notation using the proportionality sign $\propto$:

$$\phi_{wt} \propto n_{wt}, \qquad \theta_{td} \propto n_{dt}. \tag{5}$$

Equations (4), (5) define a necessary condition for a local optimum of the problem (2), (3). In the next section we will prove this for a more general case.

The system of Eqs. (4), (5) can be solved by various numerical methods. The simple iteration method leads to a family of *EM-like* algorithms, which may differ in implementation details. For example, Algorithm 2.1 avoids storing the three-dimensional array n_{dwt} by incorporating the E-step inside the M-step.

Latent Dirichlet Allocation. In LDA parameters Φ, Θ are constrained to avoid overfitting [4]. LDA assumes that the columns of the matrices Φ and Θ are drawn from the Dirichlet distributions with positive vectors of hyperparameters $\beta = (\beta_w)_{w \in W}$ and $\alpha = (\alpha_t)_{t \in T}$ respectively.

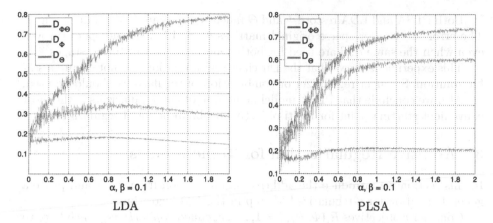

Fig. 1. Errors in restoring the matrices Φ, Θ and $\Phi\Theta$ over hyperparameter α ($\beta = 0.1$).

Learning algorithms for LDA generally fall into two categories — sampling-based algorithms [13] or variational algorithms [14]. They can be considered also as EM-like algorithms with modified M-step [15]. The following is the most simple and frequently used modification:

$$\phi_{wt} \propto n_{wt} + \beta_w, \qquad \theta_{td} \propto n_{dt} + \alpha_t. \qquad (6)$$

This modification has the effect of smoothing, since it increases small probabilities and decreases large probabilities.

The non-uniqueness problem. The likelihood (2) depends on the product $\Phi\Theta$, not on separate matrices Φ and Θ. Therefore, for any linear transformation S such that matrices $\Phi' = \Phi S$ and $\Theta' = S^{-1}\Theta$ are stochastic, their product $\Phi'\Theta' = \Phi\Theta$ gives the same value of the likelihood. The transformation S depends on a random initialization of the EM-algorithm. Thus, learning a topic model is an ill-posed problem whose solution is not unique and hence is not stable.

The following experiment on the model data verifies the ability of PLSA and LDA to restore true matrixes Φ, Θ. The collection was generated with the size parameters $|W| = 1000$, $|D| = 500$, $|T| = 30$. The lengths of the documents $n_d \in [100, 600]$ were chosen randomly. Columns of the matrices Φ, Θ were drawn from the symmetric Dirichlet distributions with parameters β, α respectively. The differences between the restored distributions $\hat{p}(i\,|\,j)$ and the model ones $p(i\,|\,j)$ were measured by the average Hellinger distance both for the matrices Φ, Θ and for their product:

$$D_\Phi = H(\hat{\Phi}, \Phi); \quad D_\Theta = H(\hat{\Theta}, \Theta); \quad D_{\Phi\Theta} = H(\hat{\Phi}\hat{\Theta}, \Phi\Theta);$$

$$H(\hat{p}, p) = \frac{1}{m} \sum_{j=1}^{m} \left(\frac{1}{2} \sum_{i=1}^{n} \left(\sqrt{\hat{p}(i\,|\,j)} - \sqrt{p(i\,|\,j)} \right)^2 \right)^{\frac{1}{2}}.$$

Both PLSA and LDA restore Φ and Θ much worse than their product, Fig. 1. The error are less for sparse original matrices Φ, Θ. LDA did not perform well even when the same α, β are used for both generating and restoring stages.

This experiment shows that the Dirichlet regularization can not ensure a stable solution. Stronger regularizer or combination of regularizers should be used.

Also we conclude that PLSA model being free of any regularizers is the most convenient starting point for multi-objective problem-oriented regularization.

3   Additive Regularization for Topic Models

In this section we introduce the additive regularization framework and prove a general equation for a regularized M-step in the EM-algorithm.

Consider r objectives $R_i(\Phi, \Theta)$, $i = 1, \dots, r$, called *regularizers*, which have to be maximized together with the likelihood (2). According to a standard scalarization approach to the multi-objective optimization we maximize a linear combination of the objectives L and R_i with nonnegative *regularization coefficients* τ_i:

$$R(\Phi, \Theta) = \sum_{i=1}^{r} \tau_i R_i(\Phi, \Theta), \qquad L(\Phi, \Theta) + R(\Phi, \Theta) \to \max_{\Phi, \Theta}. \qquad (7)$$

Topic t is called *overregularized* if $n_{wt} + \phi_{wt} \frac{\partial R}{\partial \phi_{wt}} \leq 0$ for all words $w \in W$.

Document d is called *overregularized* if $n_{dt} + \theta_{td} \frac{\partial R}{\partial \theta_{td}} \leq 0$ for all topics $t \in T$.

Theorem 1. *If the function $R(\Phi, \Theta)$ is continuously differentiable and (Φ, Θ) is the local minimum of the problem (7), (3), then for any topic t and any document d that are not overregularized the system of equations holds:*

$$n_{dwt} = n_{dw} \frac{\phi_{wt} \theta_{td}}{\sum_{s \in T} \phi_{ws} \theta_{sd}}; \qquad (8)$$

$$\phi_{wt} \propto \left(n_{wt} + \phi_{wt} \frac{\partial R}{\partial \phi_{wt}} \right)_+; \qquad n_{wt} = \sum_{d \in D} n_{dwt}; \qquad (9)$$

$$\theta_{td} \propto \left(n_{dt} + \theta_{td} \frac{\partial R}{\partial \theta_{td}} \right)_+; \qquad n_{dt} = \sum_{w \in d} n_{dwt}; \qquad (10)$$

where $(z)_+ = \max\{z, 0\}$.

Note 1. Equation (9) gives $\phi_t = 0$ for overregularized topics t. Equation (10) gives $\theta_d = 0$ for overregularized documents d. Overregularization is an important mechanism, which helps to exclude insignificant topics and documents out of the topic model. Regularizers that encourage topic exclusions may be used to optimize the number of topics. A document may be excluded if it is too short or does not contain topical words.

Note 2. The system of Eqs. (8)–(10) defines a regularized EM-algorithm. It keeps E-step from (4) and redefines M-step by regularized Eqs. (9), (10). If $R(\Phi, \Theta) = 0$ then the regularized topic model is reduced to the usual PLSA.

Proof. For the local minimum (Φ, Θ) of the problem (7), (3) the KKT conditions (see Appendix A) can be written as follows:

$$\sum_d n_{dw} \frac{\theta_{td}}{p(w\,|\,d)} + \frac{\partial R}{\partial \phi_{wt}} = \lambda_t - \lambda_{wt}; \quad \lambda_{wt} \geq 0; \quad \lambda_{wt}\phi_{wt} = 0.$$

Let us multiply both sides of the first equation by ϕ_{wt}, reveal the auxiliary variable n_{dwt} from (8) in the left-hand side and sum it over d:

$$\phi_{wt}\lambda_t = \sum_d n_{dw} \frac{\phi_{wt}\theta_{td}}{p(w\,|\,d)} + \phi_{wt}\frac{\partial R}{\partial \phi_{wt}} = n_{wt} + \phi_{wt}\frac{\partial R}{\partial \phi_{wt}}.$$

An assumption that $\lambda_t \leq 0$ contradicts the condition that topic t is not overregularized. Then $\lambda_t > 0$, $\phi_{wt} \geq 0$, the left-hand side is nonnegative, thus the right-hand side is nonnegative too, consequently,

$$\phi_{wt}\lambda_t = \left(n_{wt} + \phi_{wt}\frac{\partial R}{\partial \phi_{wt}} \right)_+. \tag{11}$$

Let us sum both sides of this equation over all $w \in W$:

$$\lambda_t = \sum_{w \in W} \left(n_{wt} + \phi_{wt}\frac{\partial R}{\partial \phi_{wt}} \right)_+. \tag{12}$$

Finally, we obtain (9) by expressing ϕ_{wt} from (11) and (12).

Equations for θ_{td} can be derived analogously thus finalizing the proof.

The EM-algorithm for learning regularized topic models can be implemented by easy modification of any EM-like algorithm at hand. In Algorithm 2.1 only steps 7 and 8 are to be modified according to Eqs. (9) and (10).

4   A Survey of Regularizers for Topic Models

In this section we revisit some of the well known topic models and show that ARTM significantly simplifies their inference and modifications. We propose an alternative interpretation of LDA as a regularizer that minimizes KL-divergence with a fixed distribution. Then we revisit topic models for sparsing domain-specific topics, smoothing background (common lexis) topics, semi-supervised learning, number of topics optimization, topics decorrelation, topic coherence maximization, documents linking, and document classification. We also consider the problem of combining regularizers and introduce the notion of *regularization trajectory*.

Smoothing regularization and LDA. Let us minimize the KL-divergence (see Appendix B) between the distributions ϕ_t and a fixed distribution $\beta = (\beta_w)_{w \in W}$, and the KL-divergence between θ_d and a fixed distribution $\alpha = (\alpha_t)_{t \in T}$:

$$\sum_{t \in T} \mathrm{KL}_w(\beta_w \| \phi_{wt}) \to \min_\Phi, \qquad \sum_{d \in D} \mathrm{KL}_t(\alpha_t \| \theta_{td}) \to \min_\Theta.$$

After summing these criteria with coefficients β_0, α_0 and removing constants we have the regularizer

$$R(\Phi, \Theta) = \beta_0 \sum_{t \in T} \sum_{w \in W} \beta_w \ln \phi_{wt} + \alpha_0 \sum_{d \in D} \sum_{t \in T} \alpha_t \ln \theta_{td} \to \max.$$

The regularized M-step (9) and (10) gives us two equations

$$\phi_{wt} \propto n_{wt} + \beta_0 \beta_w, \qquad \theta_{td} \propto n_{dt} + \alpha_0 \alpha_t,$$

which are exactly the same as the M-step (6) in LDA model with hyperparameter vectors $\beta = \beta_0(\beta_w)_{w \in W}$ and $\alpha = \alpha_0(\alpha_t)_{t \in T}$ of the Dirichlet distributions.

The non-Bayesian interpretation of the smoothing regularization in terms of KL-divergence is simple and natural. Moreover, it avoids complicated inference techniques such as Variational Bayes or Gibbs Sampling.

Sparsing regularization. The opposite regularization strategy is to maximize KL-divergence between ϕ_t, θ_d and fixed distributions β, α:

$$R(\Phi, \Theta) = -\beta_0 \sum_{t \in T} \sum_{w \in W} \beta_w \ln \phi_{wt} - \alpha_0 \sum_{d \in D} \sum_{t \in T} \alpha_t \ln \theta_{td} \to \max.$$

For example, to find a sparse distributions ϕ_{wt} with lower entropy we may choose the uniform distribution $\beta_w = \frac{1}{|W|}$, which is known to have the largest entropy.

The regularized M-step (9) and (10) gives equations that differ from the smoothing equations only in the sign of the parameters β, α:

$$\phi_{wt} \propto \left(n_{wt} - \beta_0 \beta_w\right)_+, \qquad \theta_{td} \propto \left(n_{dt} - \alpha_0 \alpha_t\right)_+.$$

The idea of entropy-based sparsing was originally proposed in the dynamic PLSA for video processing tasks [16] to produce sparse distributions of topics over time. The Dirichlet prior conflicts with sparsing assumption, which leads to sophisticated sparse LDA models [5–9]. Simple and natural sparsing is possible only by abandoning the Dirichlet prior assumption.

Combining smoothing and sparsing. In modeling a multidisciplinary text collection topics should contain domain-specific words and be free of common lexis words. To learn such a model we suggest to split the set of topics T into two subsets: sparse domain-specific topics S and smoothed background topics B. Background topics should be close to a fixed distribution over words β_w and should appear in all documents. The model with background topics B is an extension of robust models [17, 18], which used a single background distribution.

Semi-supervised learning. Additional training data can further improve quality and interpretability of a topic model. Assume that we have a prior knowledge, stating that each document d from a subset $D_0 \subseteq D$ is associated with a subset of topics $T_d \subset T$. Analogically, assume that each topic $t \in T_0$ contains a subset of words $W_t \subset W$. Consider a regularizer that maximizes the total probability of topics in T_d and the total probability of words in W_t:

$$R(\Phi, \Theta) = \beta_0 \sum_{t \in T_0} \sum_{w \in W_t} \phi_{wt} + \alpha_0 \sum_{d \in D_0} \sum_{t \in T_d} \theta_{td} \to \max.$$

The regularized M-step (9) and (10) gives yet another sort of smoothing:

$$\phi_{wt} \propto n_{wt} + \beta_0 \phi_{wt}, t \in T_0, w \in W_t; \quad \theta_{td} \propto n_{dt} + \alpha_0 \theta_{td}, d \in D_0, t \in T_d.$$

Sparsing regularization of topic probabilities for the words $p(t \mid d, w)$ is motivated by a natural assumption that each word in a text is usually related to one topic. To meet this requirement we use the entropy-based sparsing and maximize the average KL-divergence between $p(t \mid d, w)$ and uniform distribution over topics:

$$\sum_{d,w} n_{dw} \, \mathrm{KL}\big(\tfrac{1}{|T|} \, \big\| \, p(t \mid d, w)\big) \to \min_{\Phi, \Theta};$$

$$R(\Phi, \Theta) = \frac{\tau}{|T|} \sum_{d,w} n_{dw} \sum_{t \in T} \ln \frac{\sum_{s \in T} \phi_{ws} \theta_{sd}}{\phi_{wt} \theta_{td}} \to \max.$$

The regularized M-step (9) and (10) gives

$$\phi_{wt} \propto \big(n_{wt} + \tau\big(n_{wt} - \tfrac{1}{|T|} n_w\big)\big)_+, \qquad \theta_{td} \propto \big(n_{dt} + \tau\big(n_{dt} - \tfrac{1}{|T|} n_d\big)\big)_+.$$

These equations mean that ϕ_{wt} decreases (and may eventually turn to zero) if the word w occurs in the topic t less frequently than in the average over all topics. Analogously, θ_{td} decreases (and may also turn to zero) if the topic t occurs in the document d less frequently than in the average over all topics.

Elimination of insignificant topics can be done by entropy-based sparsing of the global distribution over topics $p(t) = \sum_d p(d)\theta_{td}$. To do this we maximize the KL-divergence between $p(t)$ and the uniform distribution over topics:

$$R(\Theta) = \tau \sum_{t \in T} \ln \sum_{d \in D} p(d)\theta_{td} \to \max.$$

The regularized M-step (10) gives

$$\theta_{td} \propto \Big(n_{dt} - \tau \frac{n_d}{n_t} \theta_{td}\Big)_+.$$

This regularizer works as a row sparser for the matrix Θ because of n_t counter in the denominator. If n_t is small then the big values are subtracted from all elements n_{dt} of the t-th row of the matrix Θ. If all elements of a row will be set to zero then the corresponding topic t could never be used, i.e. it will be eliminated from the model. We can decrease the current number of active topics gradually during EM-iterations by increasing a coefficient τ until some of the quality measures will not deteriorate.

Note that this approach to the number of topics optimization is much simpler than the state-of-the-art Bayesian techniques such as Hierarchical Dirichlet Process [19] and Chinese Restaurant Process [20].

Covariance regularization for topics. Reducing the overlapping between the topic-word distributions is known to make the learned topics more interpretable [21]. A regularizer that minimizes covariance between vectors ϕ_t,

$$R(\Phi) = -\tau \sum_{t \in T} \sum_{s \in T \setminus t} \sum_{w \in W} \phi_{wt} \phi_{ws} \to \max,$$

leads to the following equation of the M-step:

$$\phi_{wt} \propto \left(n_{wt} - \tau \phi_{wt} \sum_{s \in T \setminus t} \phi_{ws} \right)_+.$$

That is, for each word w the highest probabilities ϕ_{wt} will increase from iteration to iteration, while small probabilities will decrease, and may eventually turn into zeros. Therefore, this regularizer also stimulates sparsity. Besides, it has another useful property, which is to group stop-words into separate topics [21].

Covariance regularization for documents. Sometimes we possess an information that some documents are likely to share similar topics. For example, they may fall into the same category or one document may have a reference or a link to the other. Making use of this information in terms of the regularizer, we get:

$$R(\Theta) = \tau \sum_{d,c} n_{dc} \sum_{t \in T} \theta_{td} \theta_{tc} \to \max,$$

where n_{dc} is the weight of the link between documents d and c. A similar LDA-JS model is described in [22], which is based on the minimization of Jensen–Shannon divergence between θ_d and θ_c, rather than on the covariance maximization.

According to (10), the equation for θ_{td} in the M-step turns into

$$\theta_{td} \propto n_{dt} + \tau \theta_{td} \sum_{c \in D} n_{dc} \theta_{tc}.$$

Thus the iterative process adjusts probabilities θ_{td} so that they become closer to θ_{tc} for all documents c, connected with d.

Coherence maximization. A topic is called *coherent* if the most frequent words from this topic typically appear nearby in the documents (either in the training collection, or in some external corpus like Wikipedia). An average topic coherence is known to be a good measure of interpretability of a topic model [23].

Consider a regularizer, which augments probabilities of coherent words [24]:

$$R(\Phi) = \tau \sum_{t \in T} \ln \sum_{u,v \in W} C_{uv} \phi_{ut} \phi_{vt} \to \max,$$

where $C_{uv} = N_{uv} [\mathrm{PMI}(u, v) > 0]$ is the co-occurrence estimate of word pairs $(u, v) \in W^2$, pointwise mutual information $\mathrm{PMI}(u, v) = \ln \frac{|D| N_{uv}}{N_u N_v}$ is defined through document frequencies: N_{uv} is the number of documents that contain both words u, v in a sliding window of ten words, N_u is the number of documents that contain at least one occurrence of the word u.

Note that there is no common approach to the coherence optimization in the literature. Another coherence optimizer was proposed in [25] for LDA model and Gibbs Sampling algorithm with more complicated motivations through a generalized Polya urn model and a more complex heuristic estimate for C_{wv}. Again, this regularizer can be much easier reformulated in terms of ARTM.

The classification regularizer. Let C be a finite set of classes. Suppose each document d is labeled by a subset of classes $C_d \subset C$. The task is to infer a relationship between classes and topics, improve a topic model by using labels information, and to learn a decision rule to classify new documents. Common discriminative approaches such as SVM or Logistic Regression usually give unsatisfactory results on large text collections with a big number of unbalanced and interdependent classes. Probabilistic topic models can benefit in this situation [2].

Recent research papers provide various examples of document labeling. Classes may refer to text categories [2,26], authors [27], time periods [16,28], cited documents [22], cited authors [29], users of documents [30]. Many specialized models has been developed for these and other cases, more information can be found in surveys [2,3]. All these models fall into a small number of types that can be easily expressed in terms of ARTM. Below we consider one of the most general topic model for document classification.

Let us expand the probability space to the set $D \times W \times T \times C$ and assume that each word w in each document d is not only related to a topic $t \in T$, but also to a class $c \in C$. To classify documents we model a distribution $p(c \mid d)$ over classes for each document d. As in the Dependency LDA topic model [2], we assume that $p(c \mid d)$ is expressed in terms of distributions $p(c \mid t) = \psi_{ct}$ and $p(t \mid d) = \theta_{td}$ in a way, similar to the basic topic model (1):

$$p(c \mid d) = \sum_{t \in T} \psi_{ct} \theta_{td},$$

where $\Psi = (\psi_{ct})_{C \times T}$ is a new model parameters matrix. Our regularizer minimize KL-divergence between the probability model of classification $p(c \mid d)$ and the empirical frequency $m_{dc} = n_d \frac{[c \in C_d]}{|C_d|}$ of classes in the documents:

$$R(\Psi, \Theta) = \tau \sum_{d \in D} \sum_{c \in C} m_{dc} \ln \sum_{t \in T} \psi_{ct} \theta_{td} \to \max.$$

The problem is still solved via EM-like algorithms. In addition to (4), the E-step estimates conditional probabilities $p(t \mid d, c)$ and auxiliary variables m_{dct}:

$$m_{dct} = m_{dc} p(t \mid d, c), \qquad p(t \mid d, c) = \frac{\psi_{ct} \theta_{td}}{\sum_{s \in T} \psi_{cs} \theta_{sd}}.$$

In the M-step ϕ_{wt} are estimated from (5), the estimates for ψ_{ct} are analogous to ϕ_{wt}, the estimates for θ_{td} accumulate counters of words and classes within the documents:

$$\psi_{ct} \propto m_{ct}, \quad m_{ct} = \sum_{d \in D} m_{dct}; \qquad \theta_{td} \propto n_{dt} + \tau m_{dt}, \quad m_{dt} = \sum_{c \in C} m_{dct}.$$

Additional regularizers for Ψ can be used to control sparsity.

Label regularization improves classification for multi-label classification problems with unbalanced classes [2] by minimizing KL-divergence between the model distribution $p(c)$ over classes and the empirical frequencies of classes $\hat{p}_c$ observed in the training data:

$$R(\Psi) = \tau \sum_{c \in C} \hat{p}_c \ln p(c) \to \max; \qquad p(c) = \sum_{t \in T} \psi_{ct} p(t), \quad p(t) = \frac{n_t}{n}.$$

The formula for the M-step is therefore as follows:

$$\psi_{ct} \propto m_{ct} + \tau \hat{p}_c \frac{\psi_{ct} n_t}{\sum_{s \in T} \psi_{cs} n_s}.$$

Regularization trajectory. A linear combination of multiple regularizers R_i depends on regularization coefficients τ_i, which require a special handling in practice. A similar problem is efficiently solved in ElasticNet algorithm, which combines L_1 and L_2-regularizers for regression and classification tasks [31]. In topic modeling there are far more various regularizers and they can influence each other in a non-trivial way. Our experiments show that some regularizers may worsen the convergence if they are activated too early or too abruptly. Therefore our recommendation is to choose the regularization trajectory experimentally.

5   Quality Measures for Topic Models

The accuracy of a topic model $p(w \mid d)$ on the collection D is commonly evaluated in terms of *perplexity* closely related to the likelihood

$$\mathscr{P}(D, p) = \exp\left(-\frac{1}{n} L(\Phi, \Theta)\right) = \exp\left(-\frac{1}{n} \sum_{d \in D} \sum_{w \in d} n_{dw} \ln p(w \mid d)\right).$$

The *hold-out perplexity* $\mathscr{P}(D', p_D)$ of the model p_D trained on the collection D is evaluated on the test set of documents D', which does not overlap with D. In our experiments we split the collection randomly so that $|D| : |D'| = 10 : 1$. Each testing document d is further randomly split into two halves: the first one is used to estimate parameters θ_d, and the second one is used in the perplexity evaluation. The words in the second halves that did not appear in D are ignored. Parameters ϕ_t are estimated from the training set.

The *sparsity* of a model is measured by the percent of zero elements in matrices Φ and Θ. For the models that separate domain-specific topics S and background topics B we estimate sparsity over domain-specific topics S only.

The high *ratio of background words* over document collection

$$\text{BackgroundRatio} = \frac{1}{n} \sum_{d \in D} \sum_{w \in d} \sum_{t \in B} p(t \mid d, w)$$

may indicate the model degradation as a result of excessive sparsing or topics elimination and can be used as a stopping criterion for sparsing.

The *interpretability* of a topic model is evaluated indirectly by coherence, which is known to correlate well with human interpretability [23,25,32]. The *coherence of a topic* is defined as the *pointwise mutual information* averaged over all pairs of words within the k most probable words of the topic t:

$$\mathrm{PMI}_t = \frac{2}{k(k-1)} \sum_{i=1}^{k-1} \sum_{j=i}^{k} \mathrm{PMI}(w_i, w_j)$$

where w_i is the i-th word in the list of ϕ_{wt}, $w \in W$, sorted in descending order. *Coherence of a topic model* is defined as average PMI_t over all domain-specific topics $t \in S$. In most papers the value k is fixed to 10. Due to a particular importance of the topic coherence we have also examined two additional measures: the coherence for $k = 100$, and the coherence for the topic kernels.

We define the *kernel* of each topic as a set of words that distinguish this topic from other topics: $W_t = \{w \colon p(t \,|\, w) > \delta\}$. In our experiments we set $\delta = 0.25$. We suggest that well interpretable topic must have a reasonable *kernel size* $|W_t|$ about 20–200 words and a high values of topic *purity* and *contrast*:

$$\mathrm{Purity}_t = \sum_{w \in W_t} p(w \,|\, t); \qquad \mathrm{Contrast}_t = \frac{1}{|W_t|} \sum_{w \in W_t} p(t \,|\, w).$$

We define the corresponding measures of the overall topic model (kernel size, purity and contrast) by averaging over all domain-specific topics $t \in S$.

6 Experiments with Combining Regularizers

We are going to demonstrate ARTM approach in practice by combining regularizers for sparsing, smoothing, topics decorrelation, and number of topics optimization. Our objective is to build a highly sparse topic model with a better interpretability of topics, and at the same time to extract stop-words and common lexis words. Thus, we aim to improve several quality measures with no significant loss of the likelihood or perplexity.

Text collection. In our experiments we use the NIPS dataset, which contains $|D| = 1566$ English articles from the Neural Information Processing Systems conference. The length of the collection in words is $n \approx 2.3 \cdot 10^6$. The vocabulary size is $|W| \approx 1.3 \cdot 10^4$. The testing set has $|D'| = 174$ documents.

In the preparation step we used BOW toolkit [33] to perform changing to low-case, punctuation elimination, and stop-words removal.

In all the experiments the number of iterations was set to 100, and the number of topics was set to $|T| = 100$ with $|B| = 10$ background topics.

Experimental results. Figures 2–3 present quality measures of the topic model as a function of the iteration step. In each figure we compare two models, PLSA being shown with grey lines and ARTM with black lines.

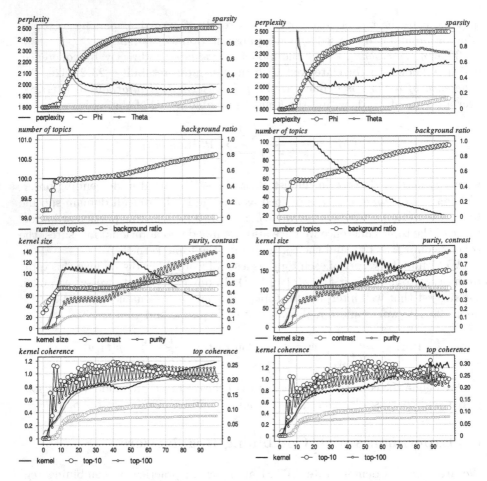

Fig. 2. Comparing PLSA (grey) vs. ARTM with sparsing, smoothing, and decorrelation (black).

Fig. 3. Comparing PLSA (grey) vs. ARTM with sparsing, smoothing, and decorrelation and topics elimination (black).

Quality measures are shown in four charts, stack on top of each other in one column with synchronized horizontal axes. Top chart: perplexity on the left-hand axis, and sparsity of matrices Φ, Θ on the right-hand axis. Second chart: number of topics on the left-hand axis, and ratio of background words on the right-hand axis. Third chart: kernel size on the left-hand axis, and contrast and purity on the right-hand axis. Bottom chart: kernel coherence on the left-hand axis, and top10 and top100 coherence on the right-hand axis.

ARTM allows to use regularizers in any combination. Therefore, we explore how various combinations of regularizer influence different quality measures.

PLSA and LDA have performed similarly by all measures: perplexity is about 1900; sparsity is 0 %; kernel size is 80–100 words; purity is 12 %; contrast is 43 %; coherence top10: 0.07, top100: 0.12, kernel: 0.9.

In ARTM we augment the regularization coefficient for sparsing gradually from the 10-th iteration. An earlier or a more abrupt sparsing may lead to perplexity deterioration. The gradual sparsing results in a highly sparse Φ matrix (98 % of zeros) and Θ matrix (85 % of zeros), while the perplexity becomes slightly worse. We smooth the background topics from the first iteration using the uniform distribution $\beta_w = 1/|W|$ and parameters $\alpha = 0.8$, $\beta = 0.1$. Using a non-uniform distribution $\beta_w = n_w/n$ yields similar results.

The decorrelation regularizer works well if activated from the very beginning. It does not change the perplexity significantly, and improves purity and coherence. Contrast and kernel size remain the same. However, the sparsity of Φ stays at 40 %, which apparently is not good enough, and Θ does not get sparse at all. The combination of sparsing, smoothing and decorrelation provides the best results, shown in Fig. 2. Notice that in all experiments kernel coherence is considerably higher than top10 and top100 coherence.

The sparsing regularizer for insignificant topics elimination turned out to be in conflict with decorrelation. Therefore we apply decorrelation at even iterations, and topics elimination at odd iterations. In our experiments the removal of topics begins to deteriorate the model perplexity when the number of topics becomes less than 60, Fig. 3.

7   Conclusions

This tutorial gives a brief survey of topic models from a new non-Bayesian viewpoint which we call ARTM — *Additive Regularization of Topic Models*. ARTM makes topic models easy to design, easy to infer, and easy to explain. Many topic models are based on stochastic matrix factorization — an ill-posed inverse problem whose solution is non-unique and instable. The goal of regularization is to reduce a potentially infinite set of solutions, and to select a better one, which satisfies our additional requirements. These requirements can be formalized through a maximization of a weighted sum of regularizers, differentiable with respect to the parameters of the model. The EM-algorithm with a modified M-step can be used to solve the optimization problem. Our interpretation of the EM-algorithm is also nonprobabilistic. We consider the EM-algorithm as a simple iteration method for solving a system of equations that defines a necessary conditions of the local optimum. Problems of a numerical convergence and regularization trajectories are left beyond the scope of this paper.

Acknowledgements. The work was supported by the Russian Foundation for Basic Research grants 14-07-00847, 14-07-00908. We thank Alexander Frey for his help and valuable discussion, and Vitaly Glushachenkov for his experimental work on model data.

A   The Karush–Kuhn–Tucker (KKT) Conditions

Consider the following nonlinear optimization problem:

$$f(x) \to \max_x; \qquad g_i(x) \geq 0, i = 1, \ldots, m; \qquad h_j(x) = 0, j = 1, \ldots, k.$$

Suppose that the objective function $f\colon \mathbb{R}^n \to \mathbb{R}$ and the constraint functions $g_i\colon \mathbb{R}^n \to \mathbb{R}$ and $h_j\colon \mathbb{R}^n \to \mathbb{R}$ are continuously differentiable at a point x^*. If x^* is a local maximum that satisfies some regularity conditions (which are always true if g_i and h_j are linear functions), then there exist constants μ_i, $i = 1, \ldots, m$ and λ_j, $j = 1, \ldots, k$, called KKT multipliers, such that

$$\frac{\partial}{\partial x}\left(f(x) + \sum_{i=1}^{m} \mu_i g_i(x) + \sum_{j=1}^{k} \lambda_j g_j(x) \right) = 0; \qquad \text{(stationarity)}$$

$$g_i(x) \geq 0; h_j(x) = 0; \qquad\qquad\qquad \text{(primal feasibility)}$$

$$\mu_i \geq 0; \qquad\qquad\qquad\qquad\qquad \text{(dual feasibility)}$$

$$\mu_i g_i(x) = 0. \qquad\qquad\qquad\qquad \text{(complementary slackness)}$$

B   The Kullback–Leibler Divergence

The Kullback–Leibler divergence or relative entropy is a non-symmetric measure of the difference between probability distributions $P = (p_i)_{i=1}^{n}$ and $Q = (q_i)_{i=1}^{n}$:

$$\mathrm{KL}(P\|Q) \equiv \mathrm{KL}_i(p_i\|q_i) = \sum_{i=1}^{n} p_i \ln \frac{p_i}{q_i}.$$

From the informational point of view, $\mathrm{KL}(P\|Q)$ is a measure of the information lost when Q is used to approximate P. KL-divergence measures the expected number of extra bits required to code samples from P when using a code based on Q, rather than using a code based on P. Typically P represents the empirical distribution of data, Q represents a model or approximation of P.

The KL-divergence is always non-negative.

$\mathrm{KL}(P\|Q) = 0$ if and only if $P = Q$.

The KL-divergence minimization is equivalent to the likelihood maximization of a model distribution $Q(\alpha)$ over parameter vector α:

$$\mathrm{KL}(P\|Q(\alpha)) = \sum_{i=1}^{n} p_i \ln \frac{p_i}{q_i(\alpha)} \to \min_{\alpha} \quad \Longleftrightarrow \quad \sum_{i=1}^{n} p_i \ln q_i(\alpha) \to \max_{\alpha}.$$

References

1. Blei, D.M.: Probabilistic topic models. Commun. ACM **55**(4), 77–84 (2012)
2. Rubin, T.N., Chambers, A., Smyth, P., Steyvers, M.: Statistical topic models for multi-label document classification. Mach. Learn. **88**(1–2), 157–208 (2012)
3. Daud, A., Li, J., Zhou, L., Muhammad, F.: Knowledge discovery through directed probabilistic topic models: a survey. Front. Comput. Sci. China **4**(2), 280–301 (2010)
4. Blei, D.M., Ng, A.Y., Jordan, M.I.: Latent Dirichlet allocation. J. Mach. Learn. Res. **3**, 993–1022 (2003)

5. Shashanka, M., Raj, B., Smaragdis, P.: Sparse overcomplete latent variable decomposition of counts data. In: Platt, J.C., Koller, D., Singer, Y., Roweis, S. (eds.) Advances in Neural Information Processing Systems, NIPS-2007, pp. 1313–1320. MIT Press, Cambridge (2008)
6. Wang, C., Blei, D.M.: Decoupling sparsity and smoothness in the discrete hierarchical dirichlet process. In: Bengio, Y., Schuurmans, D., Lafferty, J.D., Williams, C.K.I., Culotta, A. (eds.) NIPS, pp. 1982–1989. Curran Associates Inc., New York (2009)
7. Eisenstein, J., Ahmed, A., Xing, E.P.: Sparse additive generative models of text. In: ICML'11, pp. 1041–1048 (2011)
8. Larsson, M.O., Ugander, J.: A concave regularization technique for sparse mixture models. In: Shawe-Taylor, J., Zemel, R., Bartlett, P., Pereira, F., Weinberger, K. (eds.) Advances in Neural Information Processing Systems 24, pp. 1890–1898 (2011)
9. Chien, J.T., Chang, Y.L.: Bayesian sparse topic model. J. Signal Process. Syst., 1–15 (2013)
10. Khalifa, O., Corne, D.W., Chantler, M., Halley, F.: Multi-objective topic modeling. In: Purshouse, R.C., Fleming, P.J., Fonseca, C.M., Greco, S., Shaw, J. (eds.) EMO 2013. LNCS, vol. 7811, pp. 51–65. Springer, Heidelberg (2013)
11. Vorontsov, K.V.: Additive regularization for topic models of text collections. Doklady Math. 88(3) (to appear, 2014)
12. Hofmann, T.: Probabilistic latent semantic indexing. In: Proceedings of the 22nd Annual International ACM SIGIR Conference on Research and Development in Information Retrieval, pp. 50–57. ACM, New York (1999)
13. Wang, Y.: Distributed Gibbs sampling of latent dirichlet allocation: The gritty details (2008)
14. Teh, Y.W., Newman, D., Welling, M.: A collapsed variational bayesian inference algorithm for latent dirichlet allocation. In: NIPS, pp. 1353–1360 (2006)
15. Asuncion, A., Welling, M., Smyth, P., Teh, Y.W.: On smoothing and inference for topic models. In: Proceedings of the International Conference on Uncertainty in Artificial Intelligence, pp. 27–34 (2009)
16. Varadarajan, J., Emonet, R., Odobez, J.M.: A sparsity constraint for topic models – application to temporal activity mining. In: NIPS-2010 Workshop on Practical Applications of Sparse Modeling: Open Issues and New Directions (2010)
17. Chemudugunta, C., Smyth, P., Steyvers, M.: Modeling general and specific aspects of documents with a probabilistic topic model, vol. 19, pp. 241–248. MIT Press (2007)
18. Potapenko, A., Vorontsov, K.: Robust PLSA performs better than LDA. In: Serdyukov, P., Braslavski, P., Kuznetsov, S.O., Kamps, J., Rüger, S., Agichtein, E., Segalovich, I., Yilmaz, E. (eds.) ECIR 2013. LNCS, vol. 7814, pp. 784–787. Springer, Heidelberg (2013)
19. Teh, Y.W., Jordan, M.I., Beal, M.J., Blei, D.M.: Hierarchical Dirichlet processes. J. Am. Stat. Assoc. 101(476), 1566–1581 (2006)
20. Blei, D.M., Griffiths, T.L., Jordan, M.I.: The nested chinese restaurant process and bayesian nonparametric inference of topic hierarchies. J. ACM 57(2), 7:1–7:30 (2010)
21. Tan, Y., Ou, Z.: Topic-weak-correlated latent dirichlet allocation. In: 7th International Symposium Chinese Spoken Language Processing (ISCSLP), pp. 224–228 (2010)

22. Dietz, L., Bickel, S., Scheffer, T.: Unsupervised prediction of citation influences. In: Proceedings of the 24th International Conference on Machine Learning, ICML '07, pp. 233–240. ACM, New York (2007)
23. Newman, D., Noh, Y., Talley, E., Karimi, S., Baldwin, T.: Evaluating topic models for digital libraries. In: Proceedings of the 10th Annual Joint Conference on Digital Libraries, JCDL '10, pp. 215–224. ACM, New York (2010)
24. Newman, D., Bonilla, E.V., Buntine, W.L.: Improving topic coherence with regularized topic models. In: Shawe-Taylor, J., Zemel, R., Bartlett, P., Pereira, F., Weinberger, K. (eds.) Advances in Neural Information Processing Systems 24, pp. 496–504 (2011)
25. Mimno, D., Wallach, H.M., Talley, E., Leenders, M., McCallum, A.: Optimizing semantic coherence in topic models. In: Proceedings of the Conference on Empirical Methods in Natural Language Processing, EMNLP '11, pp. 262–272. Association for Computational Linguistics, Stroudsburg (2011)
26. Zhou, S., Li, K., Liu, Y.: Text categorization based on topic model. Int. J. Comput. Intell. Syst. 2(4), 398–409 (2009)
27. Rosen-Zvi, M., Griffiths, T., Steyvers, M., Smyth, P.: The author-topic model for authors and documents. In: Proceedings of the 20th Conference on Uncertainty in Artificial Intelligence, UAI '04, pp. 487–494. AUAI Press, Arlington (2004)
28. Cui, W., Liu, S., Tan, L., Shi, C., Song, Y., Gao, Z., Qu, H., Tong, X.: TextFlow: towards better understanding of evolving topics in text. IEEE Trans. Vis. Comput. Graph. 17(12), 2412–2421 (2011)
29. Kataria, S., Mitra, P., Caragea, C., Giles, C.L.: Context sensitive topic models for author influence in document networks. In: Proceedings of the Twenty-Second International Joint Conference on Artificial Intelligence, IJCAI'11, vol. 3, pp. 2274–2280. AAAI Press (2011)
30. Wang, C., Blei, D.M.: Collaborative topic modeling for recommending scientific articles. In: Proceedings of the 17th ACM SIGKDD International Conference on Knowledge Discovery and Data Mining, pp. 448–456. ACM, New York (2011)
31. Friedman, J.H., Hastie, T., Tibshirani, R.: Regularization paths for generalized linear models via coordinate descent. J. Stat. Softw. 33(1), 1–22 (2010)
32. Newman, D., Lau, J.H., Grieser, K., Baldwin, T.: Automatic evaluation of topic coherence. In: Human Language Technologies: The 2010 Annual Conference of the North American Chapter of the Association for Computational Linguistics, HLT '10, pp. 100–108. Association for Computational Linguistics, Stroudsburg (2010)
33. McCallum, A.K.: Bow: A toolkit for statistical language modeling, text retrieval, classification and clustering (1996). http://www.cs.cmu.edu/~mccallum/bow

The View on Open Data and Data Journalism: Cases, Educational Resources and Current Trends

Irina Radchenko[1(✉)] and Anna Sakoyan[2]

[1] ITMO University, St. Petersburg, Moscow, Russia
iradche@gmail.com
[2] Russian Analytical Publications Polit.ru, Moscow, Russia
ansakoy@gmail.com

Abstract. This article describes trends of open data development and a new discipline, which was formed largely due to the fact that the data have become available and open on the Internet. The authors provide a brief overview of the main directions in the development of open data and data journalism: educational projects, interaction with the community of developers using data management platforms, development of business community on open data basis. The article also discusses Russian educational projects dealing with open data and data journalism.

Keywords: Open data · Open government data · Data journalism · Open educational resources · Data expeditions · Open government

1 Introduction

There are two main features of the approach to the publication of data on the Internet in the form of open data. First, data should be freely available on the Internet, without publisher's control; second, this data must be in a reusable form. For this reason, open data should be accompanied by appropriate open licenses. It should be also submitted online in a machine-readable format. It should be machine-readable formats that allow implementing automatic processing and creating a variety of useful services based on open data.

Data journalism is one of really powerful directions within this movement [1]. The availability of data and the opportunity to create data-driven products provide a serious advantage for the informational agencies, which is specially vital now when their monopoly on information distribution is starting to crush as a consequence of the growing popularity of social networks. Moreover, the very fact of using data in media reporting is usually more explicit and visible to the audience than in many other data based products. This makes a huge contribution to the promotion of the open data movement.

D.I. Ignatov et al. (Eds.): AIST 2014, CCIS 436, pp. 47–54, 2014.
DOI: 10.1007/978-3-319-12580-0_4

2 Open Data for Education, Developers Community and Online Services

Open data have become an important tool in the development of new forms of interaction between government and civil society.

However, there is no particular value in open data, when it is not used by anyone. Open data becomes important only if there is a community of skilled users around them. One of the key components of this community is business, which comes as a stimulus for the creation of new technologies based on the new approach to open publishing data on the Internet. It is the business prospects that produced necessary conditions for the active development of data driven startups in the overseas countries, which was also encouraged by a considerable amount of governmental support at early stages of development. After all, the government is also interested in the promotion of the open data concept, because it allows for the development of new ways of dealing with civil society. This implies involving citizens in active cooperation with the government, as well as the growing transparency and accountability on the part of the authorities [2].

The practice of funding startups is widely spread in the US and UK, the two frontrunners in the area of open government and open data. This trend became even more explicit in the late 2013 and early 2014 when the federal open data portals of the US and UK changed their designs to be friendlier to their target audience of open data users. Now, there is a need for good business models of data based initiatives, as well as show cases of working and profitable projects in order to make the developers' community more motivated and the businesses more interested. It is necessary for theoretical speculations to be backed up by practical implementations, which demonstrate the prospects of using open data in analytical research and program products.

In the United States, Open Data 500 Project [3] is represented, among all, by a considerable list of startups based on open data. In the United Kingdom, The Open Data Institute has also launched a special program to support open data startups [4]. Also, steps are taken to provide the conditions for open data users to share their feedback, in order to improve the quality of the data sets that are being published. For instance, the UK federal open data portal has forums where users can discuss which directions still need more data openness, as well as the quality of the data sets that have been already published, the details of how they are published and processed, and the methods and tools for their use, etc. The portal also has blogs by the portal developers team, representatives of governmental agencies, as well as companies dealing with open data and conducting analytical research. There is also a section discussing linked open data at the portal. It provides some basic information on the topic and has a special sequence of posts tagged 'Linked Data' [5, 6]. Finally, there is a separate section on geodata, in particular the European INSPIRE Directive (Infrastructure for Spatial Information in the European Union).

In order to make working with data sets easier, developers of federal, as well as local, data portals use CKAN (The Comprehensive Knowledge Archive Network) data-publishing platform. The numerous examples include official open data portals of the UK, the US, Canada, Australia, Brazil, Spain, Slovakia, Romania, Norway, Austria, Sweden, the Netherlands, Italy, Germany, Iceland, Argentina, France, Switzerland, etc.

CKAN owes its popularity to its open-source code provided with the option of paid support by Open Knowledge Foundation.

During 2013, the popularity of CKAN was growing rapidly, as many governmental portals switched to this platform from Socrata or self-developed platforms.

Although some are still using alternative platforms, such as Socrata, Microsoft, or Koema, CKAN has already started conquering the globe. Moreover, Microsoft has provided options for CKAN deployment in its cloud (Windows Azure platform [7]).

The more portals are built using CKAN, the more comfortable it becomes for users, because it is always easier to deal with a service with a familiar interface.

It is also important that CKAN's API has single documentation [8], which means that the access to open data sets is universal and, consequently, easy to use for software developers.

In a number of states, governments went even further and openly published the source code of their services and applications, as well as documentation at GitHub [9].

Another direction, in which the open data community is developing, is educational. Apparently, it is impossible to promote the use of open data without explaining what open data is, why it is important and how to work with it.

To this end, activist communities and official institutions in different countries come up with special courses aimed at teaching the basics of working with open data. The educational direction is developed, among all, by The Open Data Institute, Open Knowledge Foundation, Knight Foundation, and Sunlight Foundation. The UN has also been interested in the promotion of the new approach. For instance, FAO UN launched a series of webinars on linked open data in late 2012 [10], and in addition to it the concept of Linked Open Data was presented in Russian as well [11]. Not long time ago, The UN launched its own online course on open data processing [12]. World Bank is also engaged in educational activities in the field of open data [13]. In September 2013 School of Open Data, organized by NGO Infoculture, was launched in Russia [14].

In some countries, a lot of efforts are aimed specifically at forming communities of software developers. These activities may include holding hackathons and organizing civil activists (The Code for America Brigades [15]), as well as launching contests for developers of open data applications. An important event in this respect is the yearly worldwide hackathon on the Open Data Day under the auspices of Open Knowledge Foundation. It should be noted that in 2014, it was the first time Russia has participated in this international event [16].

There are a lot of news subscriptions that can be used to regularly receive updates on recent open data developments. This format is actively employed by Open Knowledge Foundation, Code for America project [17], Citizens for Open Access to Civic Information and Data project [18], Map and GIS Services at the University of Toronto [19]). Special Twitter hashtags are also often used for spreading news across the community. These tags are helpful when it comes to clustering particular topics.

One more aspect of development of the open data community is building free and often online-based tools and services for data processing. An outstanding example of such products is OpenRefine, a handy tool for working with raw data. It allows cleaning datasets using different clustering methods, the method of deduplication, sorting, filtering, regular expressions and a programming language GREL (General Refine Expression Language).

It is also necessary to mention such popular tools as Google spreadsheets, Quandl [20], Raw [21], R programming language, numerous free online services for data visualization, and so on.

3 Data and Journalism

Data journalism is a specific direction in the Open Data movement, as it appears to hold an intermediary position between the data based businesses and socially explicit use of data. Like any other business at least partially based on data analysis, data journalism fits in certain profit-making business models. On the other hand, like the projects in the area of education and civil activism, it is at liberty to demonstrate the underlying mechanisms that lead to the resulting product. Moreover, unlike many spheres, in which statistics and data analysis have traditionally played a huge role (like finance or business analysis), data journalism has developed in the area traditionally associated with humanities. Although there was some amount of work with databases and numbers, still until recently there simply has not been enough material available to make it a separate profession A similar process of adapting digital techniques to traditionally non-digital spheres is now underway in many more disciplines and arts, such as literary criticism [22]. The specific of data journalism is that, unlike scholarly research, it is addressed to a vast general public audience, which potentially makes it a significant tool for promotion of open data by presenting examples of their application.

The meaning of the term data journalism seems to be generally clear, while when it comes to the details, it turns out that its application is actually defined very vaguely [23]. Even though there are numerous attempts of reflection, there is still no common understanding of what exactly it means. There are though several terms, sometimes with overlapping meanings, which are aimed to reflect certain subdirections of this trend. Here are some of them. *Data driven journalism* [24] is normally about producing a digital story which relies heavily on big data sets shown through the lens of interactive visualization. *Database journalism* [25], in its turn, is first and foremost about creating data sets and databases, which could be further used by data driven journalism. There is also *computational journalism* [26] that applies computational techniques and the elements of machine/statistical learning to creating a data driven story. A very loose term *analytic journalism* [27] might also be used in the relation to data journalism as it is sometimes based on data analysis and results in a story describing the conclusions and findings.

The obvious tendency is that more and more processes in many areas, journalism included, are becoming automated. For instance, in the recent years, there have been quite a number of examples of applying the techniques for generating human-readable texts based on data analysis to creating media articles. Several developers, including Narrative Science [28] and Automated Insights [29] have been successfully producing such tools for several years now. They may be rather expensive and not too popular among the media for now, especially if we look at the global scope, but still they are already in use there is a chance that they will become more available and therefore more powerful in the future, both in terms of the resulting product and the skills required from journalists.

That was a somewhat extreme example of digitalization applied to the traditionally human-generated content. But there are also much more common practices that have been already adopted by many leading media. Data based reporting is now becoming one of the necessary parts of a media genre system, while big newspapers create interactive visualizations or at least static infographics aimed at providing their audience with a helpful tool of exploring datasets and extracting relevant information (for examples, Kathimerini [30], New York Times [31], The Guardian [32]).

Again, the popularity of such methods seems to be different across the world at the moment. Many media companies still hesitate to invest in developing specialized data units within their staff, because hiring ready-made specialists or training the existing ones seems too expensive. Still, as the world trend to using data (including big data) in journalism is becoming stronger and more and more people are trained in working with data at least on the basic level, there is a growing need for publishers to consider the introduction of new methods. It is becoming the matter not only of fashion, but of efficiency and competitiveness as well.

In this respect, the Russian media represents an instructive example. The open data movement in the country became widely discussed only after the government resolved to start opening official data in 2013. Before that openness was mostly discussed by activists, but remained somewhat abstract and unnecessary for the majority. However, now that the data are being published, this subject evokes much broader discussion and more individuals and companies seek to make some practical use of it. Media are no exception. There are more and more examples of quality data based stories and visualizations produced by the Russian media companies (for examples, RIA Novosti [33, 34], Lenta.ru [35]).

So the process towards developing, exploring and adopting data based techniques by the media is underway. In order to be competitive, media have no choice but to become more proficient in this area. This, in turn, requires certain conditions, including: the availability of quality data, the developed tools for extracting data, the skilled specialists that can work with data and produce clear and informative data based stories, as well as visualizations. To a certain degree, it can be achieved just by the efforts of data enthusiasts who enjoy learning new techniques on their own and then apply them to their job. But this is by no means enough, unless the publishers realize the need for direct investment in this particular section.

4   Data Journalism: Educational Initiatives

Again, broadly speaking, there are some reasons to suggest that the environment necessary for making use of open data will develop naturally on its own. But it is a complex phenomenon, which is moved by a combination of top-down and bottom-up initiatives. The more people have some experience in working with data, the better they can understand the meaning and the quality of data driven stories provided by the media. The more competent the audience is, the stronger is the stimulus for the media to try hard to produce a good product. On the other hand, the less journalists are scared of learning something technical (which is a rather common fear today), the better chances they have to take interest in using data. These are two examples of what might create a healthy environment, in which the media managers would elaborate their strategies.

Speaking of the environment, one of the key factors here is broadly available, ideally open educational initiative by individual activists, not-for-profits, governments, etc. These provide not just information about what new trends exist, but also allow acquiring the basic skills, overcoming fears and understanding the meaning of what is going on. For some it may remain the only source of this kind of knowledge; others may take a more profound interest and continue learning to reach more advanced levels. All in all, educational activity in this area seems vital to maintain a healthy society.

Such initiatives have been introduced all over the world, both on national and international levels. In Russia, one of such early educational projects is DataDriven-Journalism.RU, an open resource created by a tiny team of enthusiasts in April 2013. Inspired by such internationally renowned examples as Open Knowledge foundation's School of Data [36] and Peer-to-Peer University [37], it aspires to provide a Russian-language learning opportunity. The resource as it is has a blog that discusses the relevant subjects and storage of tutorials and how-tos, both translated and original, that can be used at any time by anyone. It is also used as a central platform for so-called data-expeditions, interactive peer-learning projects aimed at developing skills of data processing and exploring certain topics. These expeditions take place online and are free of charge, so anyone who is interested can join [38]. This is an example of a bottom-up initiative that struggles to contribute to a well-balanced development of a data-driven environment.

5 Conclusion

Open Data concept is a step forward in the transformation of the Web of Documents to the Web of Data. They allow building on the basis of its new services, to conduct analytical research and empower investigative journalism, and because of this Open Data provide some freedom of information for the understanding of the processes. Open Data give new opportunities for cooperation between the government, citizens and businesses, while transforming the information environment of the Internet. In order to translate these powerful new opportunities into life, organizations (and individuals) in various countries come up with Open Educational Resources introducing the skills necessary to work with open data. Thus, new approaches and technologies become available to a wide range of stakeholders, and the information actually leads to democratization of the society and the transition to the real information society.

Acknowledgements. We take this opportunity to express our profound gratitude to many people who have helped us with their support, assistance or inspiring example. In particular, we would like to thank Ivan Begtin, Director at NGO Infoculture, for his support, encouragement and making a huge contribution to the development of the environment, in which our work takes place. We would also like to thank Lucy Chambers, Project Coordinator at OKF, whose Data MOOC at School of Data last spring was a great source of inspiration for us. Another person whose research, as well as example and encouragement have been precious for us is Vanessa Gennarelli, Learning Lead at P2P University. One of our most fruitful experiments owes a lot to the cooperation and initiative on the part of Alexey Sidorenko, the head of the project Teplitsa of Social Technologies, and his crew.

This work was partially supported by Government of Russian Federation, Grant 074-U01.

References

1. Gray, J., Chambers, L., Bounegru, L.: The Data Journalism Handbook. O'Reilly Media, Sebastopol (2012). http://datajournalismhandbook.org/
2. Ridgway, J., Smith, A.: Open data, official statistics and statistics education: threats, and opportunities for collaboration. In: Proceedings of the Joint IASE-IAOS Satellite Conference "Statistics Education for Progress", Macao, China, 22–24 August 2013
3. Open Data 500 GovLab. http://www.opendata500.com/
4. Open Data Institute. Startups. http://theodi.org/start-ups
5. Bizer, Christian, Heath, Tom, Berners-Lee, Tim: Linked data - the story so far. Int. J. Semant. Web Inf. Syst. 5(3), 1–22 (2009). doi:10.4018/jswis.2009081901
6. Hendler, J., Holm, J., Musialek, C., Thomas, G.: US Government linked open data: semantic.data.gov. IEEE Intell. Syst. 27(3), 25–31 (2012)
7. Harness Open Data with CKAN, OData and Windows Azure. http://msdn.microsoft.com/en-us/magazine/dn520247.aspx
8. CKAN. API Guide. http://docs.ckan.org/en/latest/api/index.html
9. Github and Government. https://government.github.com/community/
10. New Free Webinars @ AIMS on Linked Open Data. http://aims.fao.org/linked-open-data-webinars-at-aims
11. Sixth LOD@AIMS Webinar with Irina Radchenko on "Introduction to the Linked Open Data" (Russian). http://aims.fao.org/linked-open-data-webinars-at-aims/irina-radchenko/eng
12. Open Government Data for Citizen Engagement. http://www.unpan.org/ELearning/OnlineTrainingCentre/OpenGovernmentDataforCitizenEngagement/tabid/1751/language/en-US/Default.aspx
13. Open Data Government Toolkit. http://data.worldbank.org/open-government-data-toolkit
14. Russian Open Data School. http://opendataschool.ru/
15. The Code for America Brigade. http://brigade.codeforamerica.org/
16. Open Data day in Moscow. Open Knowledge Foundation Russia. http://ru.okfn.org/2014/02/24/odd14msk/
17. Code for America. http://codeforamerica.org/
18. Citizens for Open Access to Civic Information and Data. http://civicaccess.ca/
19. Map and GIS Services at the University of Toronto. http://mdl.library.utoronto.ca/map-gis-home
20. Quandl. http://www.quandl.com/
21. Raw. http://raw.densitydesign.org/
22. Acerbi, A., Lampos, V., Garnett, P., Bentley, R.A.: The expression of emotions in 20th century books. PLOS ONE 8(3), e59030 (2013). http://www.plosone.org/article/info:doi/10.1371/journal.pone.0059030
23. Data journalism. http://en.wikipedia.org/wiki/Data_journalism
24. Data-driven journalism. http://en.wikipedia.org/wiki/Data-driven_journalism
25. Database journalism. http://en.wikipedia.org/wiki/Database_journalism

26. Computation + Journalism. A study of Computation and Journalism and how they impact each other. http://www.computation-and-journalism.com/main/
27. Analytic journalism. http://en.wikipedia.org/wiki/Analytic_journalism
28. Narrative Science. http://narrativescience.com/
29. Let Your Data Tell Its Story. http://automatedinsights.com/
30. Kathimerini. http://www.kathimerini.gr/infographics
31. The New York Times: The Year in Interactive Storytelling (2013). http://www.nytimes.com/newsgraphics/2013/12/30/year-in-interactive-storytelling/
32. The Guardian. DataBlog. http://www.theguardian.com/news/datablog
33. RIA Novosti. Infographics. http://en.ria.ru/infographics/
34. RIA Novosti. Infographics. Russian version. http://ria.ru/infografika/
35. Island of Lost Souls. http://lenta.ru/articles/2013/09/02/dushi/
36. School of Data. http://schoolofdata.org/
37. Peer 2 Peer University. https://p2pu.org/en/
38. Sakoyan, A., Radchenko, I.: Data Expeditions and Data Journalism project as OER in Russian. http://ukwebfocus.wordpress.com/2014/03/15/data-expeditions-and-data-journalism-project-as-oer-in-russian/

Regular Papers

A Fast Mathematical Morphological Algorithm of Video-Based Moving Forklift Truck Detection in Noisy Environment

Vladimir O. Chernousov and Andrey V. Savchenko[✉]

National Research University Higher School of Economics,
Nizhniy Novgorod, Russian Federation
v.chernousov@mail.ru, avsavchenko@hse.ru

Abstract. The problem of automatic detection of the moving forklift truck in video data is explored. This task is formulated in terms of computer vision approach as a moving object detection in noisy environment. It is shown that the state-of-the-art local descriptors (SURF, SIFT, FAST, ORB) are not characterized with satisfactory detection quality if the camera resolution is low, the lighting is changed dramatically and shadows are observed. In this paper we propose to use a simple mathematical morphological algorithm to detect the presence of a cargo on the forklift truck. Its first step is the estimation of the movement direction and the front part of the truck by using the updating motion history image. The second step is the application of Canny contour detection and binary morphological operations in front of the moving object to estimate simple geometric features of empty forklift. The algorithm is implemented with the OpenCV library. Our experimental study shows that the best results are achieved if the difference of the width of bounding rectangles is used as a feature. Namely, the detection accuracy is 78.7 % (compare with 40 % achieved by the best local descriptor), while the average frame processing time is only 5 ms (compare with 35 ms for the fastest descriptor).

Keywords: Object detection · Video-based recognition · Noisy environment · Motion history image · SURF · Binary morphology · Canny operator · Forklift truck detection

1 Introduction

The relevance of the object detection task is caused by the rapid growth of the visual information flow and the need for its qualitative, rapid and complete processing [1, 2]. The presented paper is focused on a specific object detection task given by the ISS company [3], namely, the detection of a cargo on a moving forklift truck. It serves for automatic control of industrial production lines and is a typical example of the machine vision problem. By using the input video data, the real-time system should detect the forklift truck, estimate its direction, check if it contains a cargo and write the gathered information into a file for further processing.

The universal conventional solution of object detection task is implemented in such algorithms as SIFT (Scale-Invariant Feature Transform) or SURF (Speeded-Up Robust

© Springer International Publishing Switzerland 2014
D.I. Ignatov et al. (Eds.): AIST 2014, CCIS 436, pp. 57–65, 2014.
DOI: 10.1007/978-3-319-12580-0_5

Features) [4–6]. They should be applied for textured objects and are based on the keypoints detection and local descriptor extraction and comparison. However, these algorithms are highly dependent on the quality of the analyzed image and the similarity of training and test observations [2, 5]. Therefore, they show high error rate if the input data are observed in noisy environment (various pose, color of light sources, light flashing, objects partial occlusion, etc.). Unfortunately, the presence of such noise is typical in our machine vision problem.

The goal of our research is to reduce the impact of the noise (variable lighting, shadows, etc.) on the object detection accuracy if the video of the moving object is available [7] by using the mathematical morphological operations [8, 9]. In such case the moving object's silhouette may be obtained with the conventional Motion History Image (MHI) method [10]. Next, on the basis of this information about object's orientation, it is possible to compute its geometric features (width, height, are, etc.) with the mathematical morphological operations [1, 11] and, hence, estimate several attributes of analyzed object.

The rest of the paper is organized as follows: in Sect. 2, we formulate the task of the cargo detection in the moving forklift truck. In Sect. 3, we conduct an experimental study of the implemented methods and compare our results with the conventional local descriptors (SURF, SIFT, ORB, FAST). In Sect. 4, we present the findings and give concluding comments.

2   Materials and Methods

Let there be a set of $L \geq 1$ grayscale images $X_l, l = \overline{1, L}$ which defines particular object classes. Here l is the number of reference pattern in the database. The task is to assign an incoming video frame sequence $\{X(t)\}$, $t = \overline{1, T}$ holding the image of an object to one of such classes and to detect the bounding rectangle of the moving object. Here t is the frame number; T is the number of frames. This is a typical example of a video-based object detection [7]. We assume the video data that was gathered from the recorders

Fig. 1. The forklift truck with the empty fork

Fig. 2. The forklift truck with the cargo

with low resolution and has different types of noise (Gaussian, Shot, Quantization, etc.) [1, 2]. One of such practically important tasks is to detect the moving forklift truck and determine its type: is it empty (Fig. 1) or carries the cargo (Fig. 2) [3, 12].

Such a task is usually solved by detecting the keypoints (Fig. 3) and comparison of the modern local descriptors (SURF, SIFT) [4, 5]. However, this approach highly depends on the quality of the input video due to its textural basis. Therefore, a little interference can significantly worsen the detection accuracy. To test this approach we have implemented the algorithm that uses the mentioned method of moving object detection (Algorithm 1).

Fig. 3. The image of the detecting object (empty lift) with marked SURF keypoints

---

Algorithm 1. The local feature detection and comparison based method.

Data: Video frames $\{X(t)\}, t = \overline{1,T}$, empty lift image X_{trn}

Parameter: threshold of good matches M_0, matched frames threshold M_1

Result: *True* if an empty lift is presented, *False* otherwise

1. Detects key points and starts the calculation of the model image of the empty lift X_{trn}

2. Initialize the number of matched frames $N_{match} = 0$, number of frames with moving object $N_f = 0$

3. For each frame $t = \overline{1,T}$
   3.1. If the frame contains the movement, then
      3.1.1. $N_f := N_f + 1$

      3.1.2. Detects key points and calculate the descriptors of the query frame $X(t)$

      3.1.3. Compare descriptors of $X(t)$ and empty lift image X_{trn} to form good matches

      3.1.4. If number of matches exceeds threshold M_0, then
         3.1.4.1. $N_{match} := N_{match} + 1$.

4. If $N_{match} / N_f > M_1$, then return *True*

5. else return *False*

To improve the detection quality in noisy environment, we took into account the peculiarities of our task and develop the less universal, but the more accurate algorithm. The algorithm (Algorithm 2) consists of two main parts: the detection of the moving object with the MHI [10] and then determination its key-aspects with mathematical

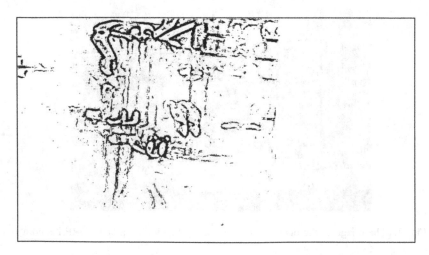

Fig. 4. The noised difference between two consequent frames

morphological operations. Note, that for the Algorithm 1, the task for motion detection is not critical because it can be extracted from the affine transform matrix taken from RANSAC [1, 4]. It takes the consecutive frame $X(t)$, compares it with the previous one X_{bg} by the MHI (Fig. 4) and gets the bounding rectangle of the moving object [8, 9], in which the contour detection operations are performed with the Canny operator [13] (Fig. 5). After that the bounding rectangles are built for large contours in the front of the moving object (obtained by MHI), among which the pair of rectangles is selected. They are considered as the bounding rectangles of an empty lift (Fig. 6).

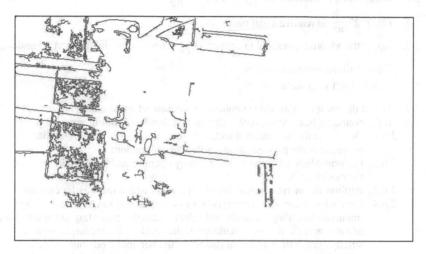

Fig. 5. The source frame with the forklift truck after contour detection with the Canny operator

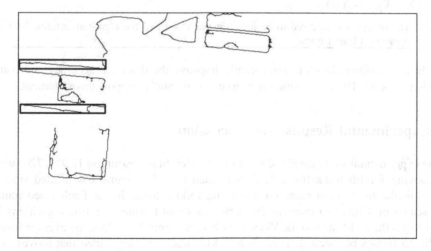

Fig. 6. The frame with the forklift truck after proposed processing with detected empty forklift (marked by black rectangles)

Algorithm 2. The mathematical morphological algorithm of empty forklift truck detection.

Data: Video frames $\{X(t)\}, t = \overline{1, T}$

Parameters: Validation criterion of the bounding region of the empty lift rectangles

Result: *True* if an empty lift is presented, *False* otherwise

1. Consider the first frame as the background image $X_{bg} = X(1)$

2. For each frame $t = \overline{2, T}$

   2.1. Background subtraction $X_{cur} = X(t) - X_{bg}$

   2.2. Filter X_{cur} to remove small noise

   2.3. Apply the adaptive threshold to convert X_{cur} into greyscale image X_g and filter it to remove the rest noise

   2.4. Update MHI with the image X_g

   2.5. Detect the contour of moving silhouette and construct its bounding box

   2.6. If the bounding box's area exceeds certain threshold (0.5 for our task), then

      2.6.1. Determine the movement direction by calculating the offset of similar rectangle in the previous frame with detected motion;

      2.6.2. Perform edges detection with the Canny operator and look for the contours in the motion area

      2.6.3. Perform the morphological closure operation with a rectangular element

      2.6.4. For each contour which corresponds to a certain size range, calculate the minimum bounding rectangle and select rectangles according to the direction of motion: if the motion is horizontal, the width of the rectangle must be strictly greater than its height (based on the horizontal position of the lift).

      2.6.5. Filter these rectangles by validation criteria. A pair of parallel rectangles witch satisfies the criteria is recognized as the empty lift

   2.7. $X_{bg} := X(t)$

3. The object is considered to be found in the video and the algorithm returns *True* if presents at least 1 frame

---

These algorithm allows to significantly improve the detection accuracy and computational speed. The next section tries to experimentally support this statement.

3   Experimental Results and Discussion

In the experimental study the ISS dataset of 22 video files (resolution 1280 × 720) with the moving forklift truck (Figs. 1, 2) was used [12]. The camera is located strictly perpendicular to the floor plane on which the forklift truck drives. Each video frame was scaled in 4 times to improve the performance of feature detection algorithms in order to get the real-time results. Video length varies from 15 to 25 s; the average frame rate is 35 frames per second. There is a forklift truck in every video that moves to a certain direction with or without the cargo. The forklift truck is the biggest moving object in the videos. There are also two types of forklift truck: a track with empty pitchforks (Fig. 1) and one with a single tip for a cargo (Fig. 2).

Both algorithms were implemented with OpenCV library [14]. In the first part of an experiment, we test the performance of the local descriptors (Algorithm 1). We use four different detecting and comparing methods types (SURF, SIFT, ORB, FAST) from the point of detection accuracy and frame processing time. Their parameters are the threshold of good matches, the resolution of the video and the image of a detected object and the first and second nearest neighbor distance ratio (DR). The parameters values of each method type are different. The source resolution of a tested video (1280×720) was resized to 800×465 to get the real-time detection results. The resolution of the training object is equal to 303×205 (Fig. 2). The tested threshold value is taken from a range $\{16, 20, 25\}$ for SURF, $\{15, 17, 19\}$ for SIFT, $\{35, 36, 40\}$ for ORB and $\{20, 24, 30\}$ for FAST. Several values of DR were tested (0.5, 0.6, 0.7 and 0.8), and the best quality was achieved for DR equal to 0.7. Other parameters of these baseline methods are set to their default values in OpenCV library as they showed the best accuracy in our experiments. All the algorithms testing results (including experimentally founded values of the optimal parameters) are presented in Table 1.

Table 1. Local feature detection algorithm quality

	SURF	SIFT	ORB	FAST
Best good matches' threshold M_0	25	19	40	30
Best matched frames threshold M_1, %	1	2	20	5
Accuracy, %	41.8	41.1	35.5	44.5
Processing time, ms	75.1	188.8	35.5	44.5

The number of true detections is quite low even with the optimal threshold values. This is primarily because the empty lifts in the selected video are not distinguished from the background (as they both have almost the same color saturation). Therefore, 40 % detection accuracy in parallel with high frame processing time is clearly insufficient to trust the results of the algorithm.

In the second part of the experimental study, we test the implementation of our morphological detection Algorithm 2. All of its parameters were tested from the point of efficiency and algorithm's performance affect during a series of tests. To estimate an optimal preprocessing filtration type, we considered the following filters: the box filter, normalized box filter, median blur and Gaussian blur [1, 2]. After selection of an appropriate filtration type, it was tested with several conventional sizes (1, 2, ..., 9) of the filter convolution matrix (mask or kernel). The minimum size of the moving region was selected from the values of 0.5 and 0.3 of the frame because the forklift truck considered being the largest moving object in the video. To eliminate the gaps in contours, the morphological closure was performed (with rectangular element of 3 rows and 3 columns). These optimal values were detected after a series of tests: the preprocessing filtration type is a median blur, the filtration strength is equal to 5 (blur kernel size), the minimum size of the moving region is equal to 0.5.

The criteria of validation of the bounding rectangles were examined from the point of their detection accuracy: minimum height of the bounding rectangle; minimum width of the bounding rectangle; maximum difference of sizes; maximum difference in the rectangles areas; maximum distance between the rectangles; maximum relative distance between the location of the upper left corners of the rectangles. We attempted to decrease its quantity by tracing their effect on detection percentage. The results are shown on Table 2.

As one can notice from this table, all the criteria except rectangles' areas difference have a serious effect on detection accuracy. The best accuracy (78.7 %) is obtained for the combination of four last criteria. The average processing time of one frame is appeared to be very fast: 5 ms. Such results are much better than provided by textural algorithms in the both accuracy and time sides.

Table 2. Dependency of object detection quality on validation criteria

Validation criterion	Minimum height and width	Difference in rectangles areas	Maximum relative distance	Height limit	Maximum width difference
Accuracy, %	21.2	21.2	58.4	72.7	78.7
Processing time, ms	4.8	4.9	4.9	5.1	5.2

4   Conclusion and Future Work

In this paper we proposed the moving forklift truck (Figs. 1, 2) detection method, which provides 38 % higher detection accuracy and much better (almost real-time [15]) computational complexity (5 ms per frame) than traditional local descriptors in a real-world noisy video dataset [12]. As a matter of fact, the state-of-the-art textured-based detectors [2, 3] are characterized with the following key issues: (1) high dependence on the variations of the light conditions; and (2) low computing efficiency if descriptors are calculated for each video frame and the complex background contains thousands of potential keypoints. Both issues are resolved in the proposed morphological algorithm (Algorithm 2). The various lighting is processed with the Canny edge detector [13] and further binary morphological operations to reduce the random noise. The performance is improved by the usage of only primitive image processing operations without need for keypoints detection and descriptor computation in the whole video frame.

The major disadvantage of our approach is the usage of task-specific information (data about geometric sizes of the analyzed object, knowledge of absence in the scene of other objects with the same size, etc.). Thus, our method is not as universal as SIFT or SURF [4, 5]. The algorithm cannot be applied with the same geometric features (Table 2) to detect various vehicles without additional features and rules.

Acknowledgements. Andrey Savchenko is supported by RSF (Russian Science Foundation) grant 14-41-00039.

References

1. Shapiro, L., Stockman, G.: Computer Vision. Prentice Hall, Upper Saddle River (2001)
2. Sonka, M., Hlavac, V., Boyle, R.: Image Processing, Analysis, and Machine Vision, 4th edn. Cengage Learning, Stamford (2014)
3. ISS (Intelligence Secure Systems). http://www.iss.ru/
4. Lowe, D.: Distinctive image features from scale-invariant keypoints. Int. J. Comput. Vision **60**(2), 91–110 (2004)
5. Bay, H., Ess, A., Tuytelaars, T., Van Gool, L.: SURF: speeded up robust features. Comput. Vis. Image Underst. **110**(3), 346–359 (2008)
6. Savchenko, A.V.: Probabilistic neural network with homogeneity testing in recognition of discrete patterns set. Neural Netw. **46**, 227–241 (2013)
7. Savchenko, A.V.: Adaptive video image recognition system using a committee machine. Opt. Memory Neural Netw. (Inf. Opt.) **21**(4), 219–226 (2012)
8. Chien, S.-Y., Ma, S.-Y., Chen, L.-G.: Efficient moving object segmentation algorithm using background registration technique. IEEE Trans. Circuits Syst. Video Technol. **12**(7), 577–586 (2002)
9. Neria, A., Colonnese, S., Russo, G., Talone, P.: Automatic moving object and background separation. Sig. Process. **66**(2), 219–232 (1998)
10. Ahad, M.A.R., Tan, J.K., Kim, H., Ishikawa, S.: Motion history image: its variants and applications. Mach. Vis. Appl. **23**(2), 255–281 (2012)
11. Najman, L., Talbot, H. (eds.): Mathematical Morphology: From Theory to Applications. Wiley-ISTE, New York (2010)
12. ISS video dataset. ftp://isstemp:isstemp@ftpsupport.iss.ru/Loaders/Video/
13. Canny, J.A.: Computational Approach to Edge Detection, pp. 679–698. IEEE Computer Society (1986)
14. OpenCV library. http://opencv.willowgarage.com/wiki/
15. Savchenko, A.V.: Directed enumeration method in image recognition. Pattern Recogn. **45**(8), 2952–2961 (2012)

Editing and Representation of Ancient Russian Semiographic Chants on the Web

Andrey Philippovich, Marina Danshina$^{(\boxtimes)}$, and Irina Golubeva$^{(\boxtimes)}$

Bauman Moscow State Technical University, Moscow, Russia
{aphilippovich,mdanshina,igolubeva}@it-claim.ru

Abstract. This paper presents a web service for input of ancient Russian manuscript into electronic form. Researchers, historians and musicians can process musical manuscripts written in Znamenny Notation via this input system. In order to represent semiographic chants, we used a special font-face, which was developed during the project "Computer Semiography". Our objective was to study features of ancient Russian musical chants and create a system for editing and playing of these chants.

Keywords: Musical information technologies · Znamenny Notation · Semiography · Ancient musical manuscripts · Representation of chants · Song patterns · Decoding semiotic systems

1 Introduction

Thousands of ancient manuscripts containing musical notation are among the monuments of the Russian culture. Most of them have been written using special musical notation systems. Znamenny Chant has been the oldest type of Russian liturgical singing since the XI century. This notion was received from the general naming used for "Znamena" or "znamia" (distinctive mark, flag, banner), due to its particular type of musical notation (also sometimes called "Kryuk", hook, or bend). This system of musical notation takes its beginning from the early Byzantine non-linear notation also called Neume Notation. Neumes and Znamena are the precursors of modern-day musical symbols.

Semiography is a system of symbolic notation. In Russian scientific tradition, semiography also refers to the direction of research of ancient musical notations, like Znamenny Chants [1, 2]. During the past 10 years, we conducted extensive research and contributed novel developments in this direction based on methods of computational linguistics and semiotics [3–7]. Our project, "Computer Semiography" [8], seeks to provide – as nearly as possible – a lossless data representation of Znamenny Chant manuscripts. We anticipate that such representation will be useful to musicologists in Russia and around the world for sharing and analysis of ancient chant manuscripts. Additionally, we foresee that such sharing could very well take place primarily over the Internet in the form of a web-based scientific information system; that is to say, a database with special services for editing and analyzing Znamenny Chants.

© Springer International Publishing Switzerland 2014
D.I. Ignatov et al. (Eds.): AIST 2014, CCIS 436, pp. 66–77, 2014.
DOI: 10.1007/978-3-319-12580-0_6

2 Features of Ancient Russian Musical Chants

The writing system of Znamenny Chants differ from the more familiar and current use of a 5-line staff (i.e. linear notation). There is no strong correspondence between the pitch of each sign and the duration of each note when playing or notating a melody. Therefore, transferring melodies from Znamenny manuscripts to that of a modern-day five-line notation is a very complex task. Difficulties associated with decoding znamenny symbols exist in the fact that znamenny letters do not adhere to conventional notation, wherein melodies are expressed by specific semiographic signs, as opposed to something else [9].

Fig. 1. Example of a Znamenny Chant.

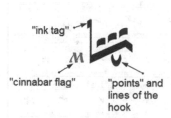

"ink tag"

"cinnabar flag"   "points" and lines of the hook

Fig. 2. Structure of a semiographic sign.

Semiographic signs possess a complex structure of main symbols and additional elements which aid a singer. Main symbols usually have a large black hook or a black stroke, several smaller black "points" and "commas", and lines near the hook or crossing through it. Additional elements – "cinnabar flag" (red color) and "ink tag" (black color) – show the pitch range of sounds and the character of its melody (Fig. 2).

Some semiographic signs may represent one, two or several notes, and some even a whole melody of more than 10 notes with a complicated rhythmic structure. The Znamenny Chant system has 12 pitches which are shared across four vocal ranges. Ink tags indicate only vocal ranges, while the cinnabar flags symbolize the highest pitch in a sequence of notes corresponding to the sign. Some cinnabar flags also indicate the nature for the notation's execution (referred to as indicatory flag).

There are several types of semiographic manuscripts, which have different structures:

- *Ordinary Chants* (Fig. 1).
- *Semiographic alphabet (primer)* – provides verbal descriptions of the semiographic signs, or matches with one or more notes (Figs. 3 and 4).
- *Collections of song patterns ("popevkas")* – different parts of a chant (stanza), which have sequences of semiographic signs that need to be translated according to special rules (Fig. 5). These patterns often have unique names and textual instructions.

- *Double noticed chants* – manuscripts (Fig. 6) which are given in direct correspondence with Linear and Znamenny Notation, and from which you can extract some knowledge regarding the translation of song patterns and semiographic signs in context, including and depending on the words with which they were matched.

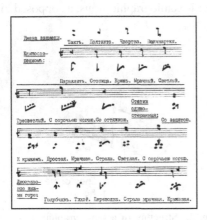

Fig. 3. Semiographic alphabet (primer) with linear notation.

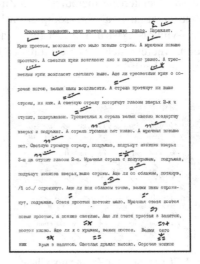

Fig. 4. Semiographic alphabet (primer) with text descriptions.

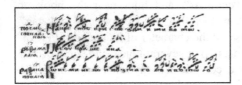

Fig. 5. Fragment of a collection of song patterns ("popevkas").

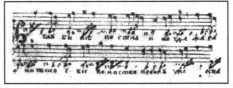

Fig. 6. Fragment of a double noticed chant ("dvoznamennik").

3   Methods for the Visualization of Znamenny Chants

Currently, a few other projects and websites share electronic versions of Znamenny Chants with internet users. One of the biggest internet resources in this field is the "Fund for Znamenny Chants".[1] This project was opened in 2003, and so far, it has managed to enter more than 10 thousand chants into Znamenny and linear notations. Developers of this project decided to convert each manuscript into separate chants. This

---

[1] "Fund for Znamenny Chants", http://znamen.ru/.

allows users to choose certain chants before viewing them; however, books cannot be viewed in their entirety by browsing from one page to the next. Additionally, chants are visualized without binding to its primary source (i.e. its scanned version).

Primary sources – more than 3,000 ancient manuscripts – are stored in state libraries, but access is restricted and only possible after obtaining a special permit. And copying a manuscript is complicated, as you have to get special permission and are only allowed to copy individual pages; not the entire manuscript. So though you are able to see how Znamenny Chants look and find information on the topic, getting access to any particular source is almost impossible for most people.

Furthermore, viewing is not the only problem. Most important is the research of manuscripts. Working with a picture or a highly fragmented book is uncomfortable because it's difficult to find information. For example, the implementation of statistical analysis takes time, as this work is quite monotonous and requires concentration. Using computers, however, does facilitate this task. There are two main abilities to represent Znamenny Chants on the Internet:

- Scan ancient manuscripts and place them on a website in graphic formats.
- Input chants in symbolic form and store them in a database with special services for viewing, searching and analyzing.

The former method is the easiest way, because it is only necessary to organize the storage of images in a specific location (e.g. a graphic database or a folder on a computer) and load them out for display to the user. There are many ready-made solutions for this, so it isn't very difficult. But this method has one main disadvantage: the inability to search through chants, as you can only look directly at pages. Additionally, more detailed queries would require special indexing.

The latter method requires developing special fonts or even software, and using them to enter chants into a text document or database. This method – unlike the former – allows the ability to search within a chant, itself; consequently, various researches could be performed.

Furthermore, the second variant takes up less space, as compared to the first. As an example, 70 pages of a book in picture format (e.g. *.bmp) could take up to 260–300 Mb, while the same page typed in a text editor might require only 2.5–3 Mb. On the other hand, a lot of effort must go into the development of fonts and to input the chants into a database. Further, it would require checking and proofreading entered chants, as people make a lot of mistakes when typing. The second method could also be accomplished in two different ways: local editing of chants using well-known software (e.g. Microsoft Word) or using special software (e.g. local or web service-based).

During this project, we tried and compared both of these variants; for which the results are represented in our publications and on the website for this project.[2] The result of our experiment: we have provided a special web service with dynamically-loadable chants and the possibility to play melodies. Users can view, edit, and play songs online. This web service is currently in development; however, initial results can already be

---

[2] Project, "Computer Semiography", http://it-claim.ru/semio.

viewed and tested. The data of this service is used in a statistical machine translation system developed during the project, called "Computer Semiography" [3–5, 7].

4   Chant Input Technology Based on a Text Processor

At the beginning of this project, we developed a special true-type font ("Andrew Semio") to import Znamenny Chants via Microsoft Word, along with the use of the "Irmologion" font set [10] for inputting the lyrics. Our Znamenny font went through several stages of improvements, ranging from experimental design to the study of an ergonomic component for this font. As a result, similar semiographic signs are located on the same letter, but can be changed with a different font style (e.g. normal, bold, and italic).

Chant input technology is shown in Fig. 7. Time expenses were for the first part ("Oktoikh") of the "Circle of Ancient Church Znamenny Chants" [11]. The scope of this book was 115 pages. The first step of this technology was to scan the raw chants. Processing a single page took about 2.5 min, with a total of 5 h to process an entire book. Importing one page of a chant takes about an hour. After importing a chant, checking for errors takes an additional 45 min. The end result is a book typeset in MS Word tables. Then the tables can be presented online and analyzed.

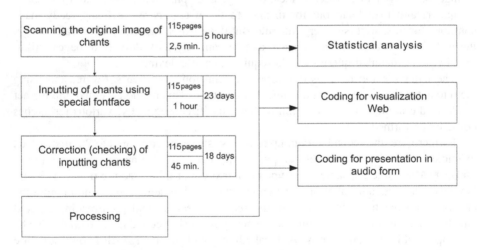

Fig. 7. Input technology for Znamenny Chants.

As a comparison, Fig. 8 shows an example of a left page from "Circle of Ancient Church Znamenny Chants" and on the right, the same page typed by using specially designed fonts. It can be observed how it is necessary to develop a font containing all ornaments, as well as a font for the title input at the first level and a font to enter the initial letters. This is necessary in order minimize the task, as well as to differentiate between the original and the electronic versions.

Fig. 8. Scanned and typed versions of a Znamenny Chant.

5 Web Services for Chant Input and Its Representation

5.1 Font Representation

There are two ways to ensure font availability. The first method would be to install the necessary fonts on a user's computer. This method is simple, but those users having additional fonts installed on their computer may be unable to perform such an installation or do it incorrectly.

The second method is an automatic replacement of server-side images and text output to a screen. There are several ready-made solutions; PCDTR (PHP + CSS Dynamic Text Replacement) [13], FLIR (Facelift Image Replacement) [18], sIFR (Scalable Inman Flash Replacement) [19]. This is a more sophisticated way; slower perhaps, but the advantage is that the user does not need to take any additional measures.

In our project for converting chants to images, we implemented it by using a specially developed technology, "PCDTR" [13]. It's a JavaScript-free version of the Dynamic Text Replacement method, originally created by Stewart Rosenberger [14]. It allows a user to take a standard HTML web page and dynamically create images to replace and enhance page headings using only PHP and CSS. These scripts have been adapted to this problem and can successfully be used as a replacement technique.

The development of fonts and the use of them for visualization purposes would allow conducting various research projects; and in the future, this may help towards decoding Znamenny Chants. Therefore, after a stage for the correction of entered chants, it will be necessary to have the most authentic data; ideally, faultless input and a minimum difference from the original document. For this purpose, it is necessary not only to improve the fonts, but also the technology. We offer the following options:

- MS Word templates.
- OCR software (Optical Character Recognition) [12].
- Other third-party programs.

5.2 Coding Semiographic Signs

The structure of ancient musical manuscripts significantly affects the coding methods data structure, coding methods and features of editing and representation services.

The first variant for storing Znamenny Chants was developed through the use of an XML format. Each sign is encoded using special attributes, utilizing the form shown in Fig. 9:

```
▼<ROWDATA>
    <ROW Znam="a" Slog="ко" Stil=" обычный" VPom="м" DPom=""/>
    <ROW Znam="Ар" Slog="кэч" Stil=" обычный Italic" VPom="в" DPom=""/>
    <ROW Znam="a" Slog="но" Stil=" Bold" VPom="п" DPom=""/>
    <ROW Znam="a" Slog="му" Stil=" Bold" VPom="п" DPom=""/>
    <ROW Znam="a" Slog="т" Stil=" Bold" VPom="п" DPom=""/>
    <ROW Znam="a" Slog="от" Stil=" Bold" VPom="п" DPom=""/>
    <ROW Znam="a" Slog="шу" Stil=" Bold" VPom="п" DPom=""/>
    <ROW Znam="a" Slog="ся" Stil=" Italic" VPom="п" DPom=""/>
```

Fig. 9. Fragment of an XML file.

Because this is unreadable, it was recoded as an original hymn using a special conversion table (Fig. 10).

As a result, we obtained a sequence of identifiers which could be applied to the translation rules in linear notation. First, the longest rules are applied, and then the shorter ones. Finally, rules from the alphabet are applied, so as not to replace semiographic signs. Thus, from the original chant notation, signs are recoded into linear notation. To play back the chant's two codes which were generated, music and time codes are stored in a separate file and read by a Flash application.

5.3 Editing and Playing Chants

A user can review a chant, compare it with its translated linear notation version and conduct visual analysis, as well as analysis by listening to its melody. Also, the user can edit the alphabet and reconvert a chant (Figs. 11 and 12).

The user selects from a list and clicks on the chant "Send" button. Afterwards, the screen displays a Znamenny Chant notation. When you click on "Transcode", the chant is converted from a semiographic notation into a linear one. The result is then shown on the screen and its melody code is generated. After displaying the original and translated chants, the matching strings are stored, which is convenient for later comparison. In order to listen to the resulting melody, the user must click on the "Play" button.

On the left part of the screen, there is a block of links to the basic data. After clicking on a link, the user is provided with a view of the alphabets and the initial chant which was created.

Fig. 10. Fragment of the coding table

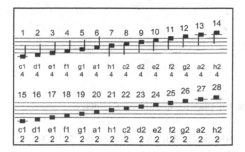

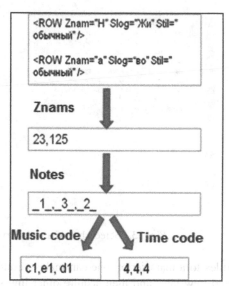

Fig. 11. Example of a code identifier matching notes.

Fig. 12. Scheme of a chant conversion.

A user can review a chant, compare it with its translated linear notation version, and conduct visual analysis, as well as analysis by listening to its melody. The user can also edit the alphabet and reconvert the chant. A screenshot of the application is shown in Fig. 13. On the left part of the screen, there is a block of links to basic data. Clicking on a link, the user is provided a view of the alphabets, as well as the initial chant which was created.

Despite the fact that a developed music editor can encode songs from Znamenny into linear notation, a system of replacement rules still hasn't been worked out; namely, an alphabet translation. The current version of the alphabet is a test version built on the alphabet of a double noticed chant, ("dvoznamennik") "Tikhon – Macarius", and is not yet complete. In order to develop a high quality alphabet, studies should be undertaken with other double noticed chant manuscripts and collections of songs patterns ("popevkas"). Additionally, manuscripts are needed to be submitted in electronic form and entered into a database; whereby, the need for a special editor for this purpose would still be necessary.

The goal of our project's development is not only to obtain a system of transfer rules, but also to obtain new knowledge about deciphering chants. Applying these new

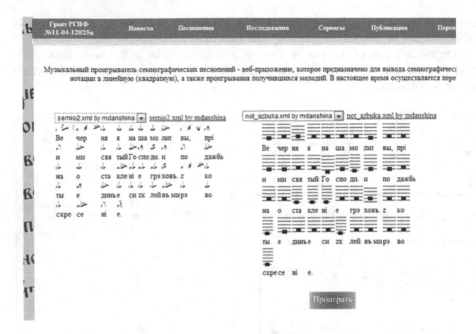

Fig. 13. Screenshot of a web service for playing Znamenny Chants.

rules to a manuscript, one can make a conclusion about the completeness of the rules and new laws and then acquire other important information.

While developing this service, the usability was also taken into account. Even though the Internet *is* widespread, it is not always available. And though the storage and viewing of manuscripts in MS Word is more convenient for viewing and listing, MS Word does not allow the processing of chants. Therefore, it is important to provide interaction both in offline *and* online operating modes (Fig. 14). The user can type a manuscript in Word, and by pressing a button, send this material to a database. To the

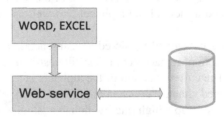

Fig. 14. Scheme of interaction for offline and online modes.

contrary, utilizing a database by means of a web application, it is then necessary to unload pages in MS Word.

5.4   Service for Inputting Znamenny Chants into a Database

For inputting chants into a database, a special program was developed. First, the user selects a manuscript book that he is going to import, then a particular page is opened. For convenience, a page of manuscript is shown on the screen as an image during the chant's input process [15].

To input a particular semiographic sign ("znamia"), users are asked to select a group to which the sign matches, and then a particular "znamia". Division into groups is conditional: semiographic signs are grouped by similarity to their outline. The program is allocated a total of 212 banners that are divided into 6 groups; each of which has one or more subgroups. The maximum number of subgroups is 7.

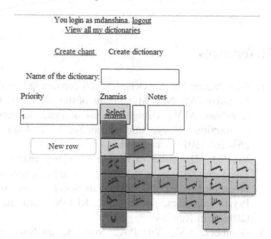

Fig. 15. Scheme example of the application for inputting chants.

Fig. 16. Scheme example of the application for inputting a translation dictionary.

By means of this application, a registered user can not only input chants, but also a dictionary containing translation rules, which can then be used further in the musical editor. During the input of a chant, the user has to choose a separate znamia and specify a syllable which corresponds to it (Fig. 15). And should a dictionary be used/applied, the user would need to choose one or even several banners and specify to what notes they correspond (Fig. 16).

6   Conclusion

Our developed system is among the first in the field of Semiography and Znamenny Chants; as such, with the exception of Russia, there are no other like-systems with which to make a comparison. In other words, to the best of our knowledge, our system has no analogies. However, issues of visualization and the processing of ancient manuscripts on the web are relevant to other domains [16, 17].

We developed and compared several methods for the representation of ancient Russian musical manuscripts and took into consideration features of chants, as well as creation of a web service for inputting Znamenny Chants. This permits the input of chants into a database and performs a faster statistical analysis of different types of manuscripts.

As a result, the database contains 29,376 records from "Circle of Ancient Church Znamenny Chant", 234 records of the appendix for "Circle of Ancient Church Znamenny Chants", 10,897 records of double noticed chants (dvoyeznamennik) and 16,914 records of the manuscript with a collection of song patterns (popevkas).

In order to listen to a resulting melody, a new web service has been developed. Any registered user can create their own semiographic alphabet with rules for decoding znamenny signs (hooks) and their sequences (patterns).

References

1. Preobrazhenskij, A.: Ocherk istorii cerkovnogo penija v Rossii (1915) - in Russian
2. Brajnikov, M.: New Monuments of the Znamenny Chant, Leningrad (1967) - in Russian
3. Danshina, M.V.: Using methods of machine translation for analysis of ancient music manuscripts. Novye informatsionnye tekhnologii v avtomatizirovannykh sistemakh 16, 263–266 (2013) - in Russian
4. Yu, P.A., Danshina, M.V.: The frequency analysis of verbal and melodic associations of Znamenny Chants. In: Challenges of Information Society and Applied Psycholinguistics - Proceedings of the X International Congress of the International Society of Psycholinguistics, pp. 251–252. RUDN - Institute of Linguistics RAN - MIL, Moscow (2013) - in Russian
5. Golubeva, I.V., Yu, P.A.: Syntactic analysis musical texts. Novye informatsionnye tekhnologii v avtomatizirovannykh sistemakh 16, 257–262 (2013) - in Russian
6. Danshina, M.V.: The method of selection, storage and processing of popevkas on musical manuscripts. In: Information Technology and Textual Heritage. materialy IV mezhdunar. nauch. konf., Petrozavodsk, 3–8 Sentyabrya 2012, pp. 79–81 (2012) - in Russian
7. Yu., P.A., Golubeva, I.V.: Research of syntax of semiographic chants. In: Proceedings of the Higher Education Institutions. Problems printing and publishing, vol. 6, pp. 147–163 (2012) - in Russian
8. Yu, P.A., Smoliakov, B.G.: Computer semiography. In: Kniga i mirovaya tsivilizatsiya: Materialy XI Mezhdunar. nauch. konf. po problemam knigovedeniya, Moskva, 20–21 April 2004, vol. 1, pp. 398–401. Nauka, Moscow (2004) - in Russian
9. Smoliakov, B.G.: About melodic ambiguity in Russian bezlineynoy stolpovoy notation. In: Sergievskie chteniya -1. O russkoy muzyke : nauch. tr. Sb., vol. 4, pp. 33–40. MGK im. P.I. Chaykovskogo, Moscow (1993) - in Russian
10. Irmologion fontset. http://en.wikipedia.org/wiki/Irmologion_(typeface)
11. Krug tserkovnago drevnyago znamennago peniya v shesti chastyakh. Obshchestvo lyubiteley drevney pis'mennosti, Saint-Petersburg (1884) - in Russian
12. Vylomova, E.A.: Recognition System of semiographic chants. In: Intellektual'nye tekhnologii i sistemy. Sbornik uchebno-metodicheskikh rabot i statey aspirantov i studentov, Moscow, vol. 9, pp. 58–70. NOK «CLAIM» (2007) - in Russian

13. Technology PCDTR. http://www.artypapers.com/csshelppile/pcdtr/
14. Rosenberger, S.: Dynamic Text Replacement. http://alistapart.com/article/dynatext
15. Danshina, M.V.: IPSM - application for inputting and processing semiographic chants. In: Information Technology and Textual Heritage. materialy III mezhdunar. nauch. konf., Izhevsk, 3–8 Sentyabrya 2010, pp. 79–82 (2010) - in Russian
16. Doumat, R., Egyed-Zsigmond, E., Pinon, J.-M.: A web 2.0 archive to access, annotate and retrieve manuscripts. Multimed. Tools Appl. **64**(1), 97–117 (2013)
17. Doumat, R., Egyed-Zsigmond, E., Pinon, J.-M.: Digitized ancient documents what's next. Document Numérique **12**(1), 31–51 (2009)
18. Facelift Image Replacement. http://mawhorter.net/flir
19. sIFR 2.0: Rich Accessible Typography for the Masses. http://www.mikeindustries.com/blog/sifr/

Detection of Domain-Specific Trends in Text Collections

Ilnur Gadelshin, Anna Antonova, and Dmitry Ilvovsky[✉]

Higher School of Economics, Moscow, Russia
{ifgadelshin, ayuantonova}@edu.hse.ru,
dilvovsky@hse.ru

Abstract. This study considers the problem of automatic trend detection in document collections related to several specific domains. The suggested trend detection algorithm is based on the domain-specific trend model. The algorithm was evaluated on documents from shipbuilding and power engineering domains.

Keywords: Trend detection · Text analysis · Document collections

1 Introduction

In this study we consider the problem of automatic trend detection in document collections, each related to a specific domain. We use sets of so called "query keywords" (QKWs) compiled by experts and assume that these collocations denote "trends" that are located closely to some QKWs in the text.

Generally, the problem is the follows. Given several document collections, each describes a particular domain at some time span, and lists of QKWs, we need to provide the sorted list of keywords and collocations, which may correspond to some trend observed in the domain at this point of time. Every collection has two labels: the domain (e.g., "shipbuilding") and the aspect (e.g., "technology", "business"). Every resulting list of keywords characterizes only one particular domain, regardless of the "aspects".

The paper is organized as follows. Section 2 provides a very brief survey of the approaches to trend detection problem. In Sects. 3 and 4 we introduce the idea of our method as well as some technical details. Then we describe a simple application of our model. Section 5 concludes the article.

2 Approaches to Trend Detection

2.1 Topic Models

One of the most popular approaches to trend detection in text collections is based on adding a time component to topic models (which are usually probabilistic). A trend is considered as a topic that changes in time. The topic model of text collection identifies

© Springer International Publishing Switzerland 2014
D.I. Ignatov et al. (Eds.): AIST 2014, CCIS 436, pp. 78–84, 2014.
DOI: 10.1007/978-3-319-12580-0_7

the topics which the document is related to and which words (terms) form each topic. The number of topics is the major parameter of the model and commonly is unknown. Therefore, the topic model is the result of joint clustering of texts and terms over topic clusters.

The paper [4] proposes the latent Dirichlet allocation (LDA) model, which is considered to be the state-of-the-art in topic modeling. LDA eliminates all weak spots of other models by using Bayesian regularization, which is an assumption that parameter vectors are generated by conjugate prior distribution. The work [5] proposes more general approach – a robust model, which exploits individual and common features of documents in collections.

The substitution of Dirichlet distribution for stochastic Dirichlet process allows us to take into account the evolution aspect within a text collection. It should be noticed that the main assumptions for topic model do not change. Some specific realizations of topic models could be found in [6–8].

Here are some disadvantages of topic models in application to trend detection problem:

1. "Bag of words" model doesn't allow us to use the structure of document;
2. Considering trend as a set of terms with probabilities of belonging to some topic conflicts with the intuitive definition of trend as a collocation;
3. The necessity of preliminary model tuning and setting the number of topics.

2.2 Keyword Tagging

Another trend analysis approach, which is used partially in our research, is based on keywords and collocations tagging. For each period of time keywords or collocations are picked out for expert examination and for further time-dependent analysis. The task is to find an intersection (full or partial) and to plot the trend evolution based on the word rank in the lists at the specific points of time. An interesting application of this approach is given in [9].

2.3 Structural and Genre-Specific Approaches

A good example of this group of methods can be found in [6]. The authors present an approach based on patent structure that uses probabilistic apparatus similar to topic models. In terms of this method a trend is defined as a sequence of technologies. The technology consists of three components: a problem, the solution of the problem (a method, an algorithm or an approach) and a time tag. Different components of technology are searched in different parts of the patent.

An obvious flaw of this group of methods is their binding to a specific field, but it is partially compensated for by the efficiency of these methods.

3 Algorithm for Domain-Specific Trend Detection in Text Collections

3.1 Domain-Specific Trend Model

The trend model considered in this work is based on the following main assumptions:

1. The trend is the only one collocation. The length of this collocation does not exceed 3 words.
2. We consider collocation as a "trend" ("technological trend") if it meets following conditions:

   (a) it is specific for the domain;
   (b) it is relevant to the documents from the given domain;
   (c) it is located close to the query keywords given by experts for this domain.

Within this paradigm, extracting collocations in text collections at some fixed moment of time allows us to build the list of potential trends. The picture of trend evolution could be obtained by using time-specific analysis of the results for every time span. However, in this study we do not use the time axis. Adding evolution in time feature is the subject of our following research.

3.2 Scheme of the Algorithm

1. Text preparation.

   (a) Removing noise symbols (not in [12–102] range of UTF-8 table), e-mails, hyperlinks.
   (b) Removing stop words (the list of stop words is taken from Python NLTK [3]).
   (c) Processing stemming (the Porter algorithm [1] implemented in Python NLTK).

2. Extracting the set of query keywords $Q = \{q_1, \ldots, q_n\}$.

   (a) Finding all the QKWs in the text, storing their positions and the number of occurrences;
   (b) Calculating the weight of each document via a simple formula:

$$weight(d) = \frac{\#\{w \in Q | w \in d\}}{\#\{w \in d\}}$$

   Weight is the fraction of query words in the document d.

3. "Meaningful" collocation extraction.

(a) Extracting all bi- and trigrams from all documents using *likelihood-ratio* measure;

(b) Withdrawing uninformative ones.

4. Twofold collocation ranking (described below).
5. Intersection of contrasting collections (described below).
6. (Optional) Reverse stemming. This step is added for making expert analysis more comfortable.

(a) Gathering all unigrams, bi- and tri-grams into the initial set S.

(b) For each stem finding the most frequent element s in the set S that contains this stem.

3.3   Collocation Ranking

In our study we introduce a twofold ranking method. The first part concerns the relative frequency of each collocation in a single document vs. cumulative frequency in all the documents in all collections. Bi- and tri-grams are ranked separately. This approach exploits a well-known *tf*idf* formula [2]:

$$w_{tfidf}(t_i) = \frac{\#\{t_i \in d_j\}}{\sum_k \#\{d \in D, D \subseteq C_k | t_i \in d\}}$$

We sort the list of terms by this value in descending order and take *top n* elements from the sorted list (we empirically set $n = 60$), after that we assign the *rank = n* to the first element in the list, the *rank = n−1* to the second element and so on. Then we calculate the cumulative rank through the whole collection of documents simply summing the values of ranks in every document. This value is used further in the formula for resulting ranking. Next we rank the term using the procedure that was already mentioned: the closer terms stay to query words, the better they are. Again for each document the *distance* measure is calculated in the following way;

$$distance(t_i) = \left(\sum_{\{q \in Q | dist(q,t_i) \leq c\}} \frac{weight(q)}{dist(q,t_i)^\alpha} \right) * \left(\frac{\#\{q_i \in Q\}}{\#\{q_i \in Q | q \in d\}} \right)$$

Intuitively, this metric describes how the particular term is close to QKWs within a specified (by parameter c) window of words. Here *dist (q, ti)* is the number of words between query word q and term t_i; $c = max(dist(q, t_i))$ is the predefined value (for evaluation we chose empirically $c = 30$); $\alpha \in [1,2]$ is a parameter defined by an expert. In the second multiplier the numerator stands for the number of QKWs within the span defined by parameter c and the denominator determines the number of QKWs in the current document. If the term t_i occurs several times in a document, the mean value of *distance(t_i)* should be taken. The next step is sorting the list of terms by *distance(t)*, taking *top-n* elements and assigning ranks to them in descending order. The rank of each term in the whole collection is the sum of ranks in every document multiplied by the document weight:

$$R_{distance}(t_i) = \sum_{d \in D} weight(d) * distance(t_i)$$

After calculating different sets of ranks, the last step is taking their linear combination:

$$R = (1 - \beta) * R_{tfidf} + \beta * R_{distance}, \beta \in [0, 1]$$

3.4 Intersection of Contrasting Collections

So far, we have ranked lists of terms for each collection. The last step depends on a particular dataset. For evaluation we used two domains and two aspects which gives us four collections: *power engineering* (business (BP) and technology (TP)) and *shipbuilding* (business (BS) and technology (TS) aspects). Collections from one domain may be exploited as a contrast to another and serve for additional ranking and/or cleaning the list of terms.

So, the final lists of terms for two domains are constructed as follows:

$$Terms_{power\_engineering} = (TP \cup BP) \ ((TS \cup BS) \cup (TS \cup TP) \cup (BS \cup BP))$$

$$Terms_{shipbuilding} = (TS \cup BS) \ ((TP \cup BP) \cup (TS \cup TP) \cup (BS \cup BP))$$

After that we make a list of *top-n* terms from each of the sorted lists of the final result. Despite the fact that using collections from contrasting domains is an essential part of our approach, we certainly shouldn't stick neither to the pair of domains (shipbuilding and power engineering) nor to the number of domains.

3.5 Implementation of the Algorithm

The proposed method was implemented in Python.[1] The application has the log-file fixing all the steps and the special ini-file where all settings and parameters could be changed. The input data for the application consists of text files and query words. The output data includes text files containing lists of ranked collocations.

4 Evaluation

As it was mentioned above, we used four collections of texts in English for testing the method. Each collection contains about 500 documents; an average document includes

---

[1] The code, instructions and results can be found on https://bitbucket.org/ilnurgadelshin/trends.

Table 1. Evaluation results for several best combinations of *alpha* and *beta*

Top-N	Junk, %		Potential trends, %	
	Ships	Power	Ships	Power
50	52	10	12	48
100	52	11	10	29
150	50	15.3	8	22
200	49	17.5	7	21

450 words. It is known that texts on energy are more genre-specific and more unified. The algorithm ran for increasing values of the final list: 50, 100, 150, 200. For each value we applied different parameters *alpha* and *beta* (Table 1).

The results of each run were evaluated by experts. Besides the trends, junk phrases were found, i.e. phrases that could be detected as not-trend by an average person: idioms, phrases on common topics, etc. The study showed that changing the parameter *alpha* has a little effect on the final result and leads only to a slight reordering of items. Decreasing the *beta* parameter to 0.5 impairs the results – it increases the amount of junk phrases and reduces the number of potential trends.

Firstly we see the increasing speed of precision reduction while increasing the length of final lists. Therefore the achieved ranking is rather good and the approximation of the recall value is high enough. Secondly, the precision itself is quite low, and that means we should improve our model. Thirdly, there is a huge difference between absolute results for two domains because of the collection quality. Fourthly, the quality of collection is very important, so the precision depends strongly on the collection. Some interesting trends for shipbuilding domain are *propulsion system*, *tunnel thrusters*, for power engineering: *wind farm, fossil fuels, and renewable energy sources.*[2]

5   Conclusions

The proposed method is applicable to other languages as well, although the language-specific tools, such as a tokenizer or a stemmer, should be replaced with those of the corresponding language. The result depends on the input, namely, genre homogeneity, topic versatility, query keywords set and weights etc. Our future work includes the exploitation of time series analysis that would help us to extract trends in the primordial meaning of this word. We also plan numerous modifications and elaboration of the existing algorithm, including preliminary clustering and developing special techniques for working with the dynamics of terms occurrence.

---

[2] The full lists of the found trends could be found on https://bitbucket.org/ilnurgadelshin/trends.

References

1. Porter, F.: An algorithm for suffix stripping. Program **14**(3), 130–137 (1980)
2. Salton, G., Buckley, C.: Term-weighting approaches in automatic text retrieval. Inf. Process. Manage. **24**(5), 513–523 (1988)
3. Natural Language Toolkit. http://www.nltk.org/
4. Blei, D.M., Ng, A.Y., Jordan, M.I.: Latent Dirichlet allocation. J. Mach. Learn. Res. **3**, 993–1022 (2003)
5. Vorontsov, K.V., Potapenko, A.A.: Regularization, robustness and sparseness of probabilistic topic models. Comput. Res. Model. **2**, 161–174 (2012) (in Russian)
6. Teh, Y.W., Jordan, M.I.: Hierarchical Bayesian nonparametric models with applications. In: Hjort, N., Holmes, C., Müller, P., Walker, S. (eds.) Bayesian Nonparametrics Principles and Practice. Cambridge University Press, Cambridge (2009)
7. Teh, Y.W.: Dirichlet processes. In: Sammut, C., Webb, G.I. (eds.) Encyclopedia of Machine Learning, pp. 280–287. Springer, Heidelberg (2010)
8. Blei, D, Lafferty, J.: Dynamic topic models. In: ICML (2006)
9. Glance, N., Hurst, M., Tomokiyo, T.: BlogPulse: automated trend discovery for weblogs. In: WWW 2004, ACM (2004)

A Study of Formal and Informal Relations of Russian-Speaking Facebook Users

Dmitry Gubanov[✉]

Digital Society Laboratory LLC,
Trapeznikov Institute of Control Sciences RAS, Moscow, Russia
dmitry.a.g@gmail.com

Abstract. The paper analyses formal relations ("friendship") among the Facebook users and studies their interrelation with informal relations ("commenting"). We define general characteristics of the network, give a definition to strong friendship links. Then we analyze strength of Facebook friendship links, study connectivity between a user's friends, and give an answer to the question how users who have something in common are connected with each other. In conclusion, we consider interrelation between friendship and commenting links.

Keywords: Social network · Formal and informal relations · Strong and weak ties

1 Introduction

Although certain social and network research was held as early as in 1930s [4], the term "social network" was introduced by sociologist James Barnes in 1954 [3]. Today a social network is interpreted, first, as a social structure comprising a set of nodes (individuals) and a set of relation links (friendship, communication etc.) defined on the first set and, second, as Internet implementation of this social structure.

Links among nodes can be interpreted differently, i.e. in fact, we can speak about of different social networks with the same set of nodes. In particular, links may be strong (e.g., regular correspondence) and weak (e.g., message exchange once a year) [5, 6], formal and informal. Formal links presuppose one-time fixation, while informal ones are supposed to be confirmed repeatedly.

The paper thoroughly considers formal "friendship" links and partly on informal "commenting" links – both in strong and weak variants. Based on the empirical data analyses an answer to the following question is being searched for: whether informal links are conditioned by formal ones (or whether formal friendship link exist by themselves and user communication is executed through other channels). The answer to this question seems to be important to solve various problems, for instance, to predict the existence of formal/informal links among users of a social network or to study the information propagation through social network links (our motivation). Unfortunately, studies on this issue have not been found (let alone studies on Russian social networks).

The structure of the paper is follows. In Sect. 1, we study friendships network: we consider general characteristics, analyze strength of Facebook friendship links, then

© Springer International Publishing Switzerland 2014
D.I. Ignatov et al. (Eds.): AIST 2014, CCIS 436, pp. 85–90, 2014.
DOI: 10.1007/978-3-319-12580-0_8

study the connectivity of the user friends and give the answer to the question how somehow similar users are connected with each other. The Sect. 2 considers the issues of formal links conditionality by informal ones and vice versa.

Anonymized dataset on Russian-speaking Facebook segment is provided for research by Digital Society Laboratory LLC[1]. According to our collaborator at Digital Society Laboratory, the dataset was collected using the API of Facebook. Commenting ties were considered from June 2012 till June 2013, friendship relations were regarded within September 2013.

2   Friendships Network

2.1   Basic Characteristics of the Friendships Network

Let us draw friendship network basic characteristics. The number of Russian speaking Facebook segment users makes 3.3 million (3,279,156), the number of friendship ties among them is equal to 77.6 million (77,639,757), a user has 47 friends on the average.

The distribution of friends number is similar to power law (alpha = 2.24): 20 % of users have not more than 3 friends, 80 % of users have not more than 45 friends.

One largest connected component (about 3.1 million users) prevails in the distribution of friendship network connected components; isolates (components consisting of a single user) are most widely spread (about 197 000 users), the number of other components does not exceed 2 dozens of users and is found considerably less often.

2.2   Friendship Weak and Strong Relations

Let us define friendship strong link. If users happen to have a common friend, friendship link between the users is called a strong one (nonrandom). Then friendship link strength of a user will be defined as a share of friends having at least one common friend with u ($|\cdot|$ means cardinality of a set):

$$w^{sf}(u) = \sum_{v \in friends(u)} I^{sf}(u, v) \Big/ |friends(u)|'$$

where $I^{sf}(u, v) = \begin{cases} 1, & \text{if there is a strong friendship link between } u \text{ and } v \\ 0, & \text{otherwise} \end{cases}$.

In general, it should be noted that the more friends a user has, the stronger (less random) friendship links he possesses (they are proved by common friends).

2.3   Links Between Friends of a User

Do friends of a certain user have friendly relations? To answer this question we use two metrics: (a) connectivity of the user's friends; (b) the number of connected components in the set of the user's friends.

---

[1] www.digsolab.com

Connectivity of the user's friends u is calculated as follows: $c^d(u) = \frac{2 \cdot ef}{d(d-1)}$, where ef is a number of links among the user's friends u, and d is the number of friends of the user u. Measure $c^d(u)$ takes the value of 0 if the friends are not somehow linked between each other, and takes the value of 1 if each friend is linked with each friend.

It is typical for a user to have 10 % friendship links between his friends from their maximal possible number. In general, for the friendship network friends connectivity of users has the value of 0.2.

The number of connected components in the set of a user's friends. On the other hand, the link of the user's friends u, $c^{wc}(u)$ can be calculated singling out the connected components in the network of the user's friends.

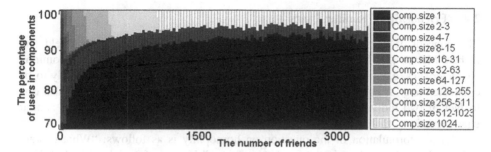

Fig. 1. Percentage of connected components of a given size for users with a given number of friends (the size of the component is the number of users in the component).

Figure 1 shows that as the number of friends in the user's neighborhood is increasing, the isolates and the largest components are dominating. However, the percentage of the isolates is higher. But. Will these ratios be observed if we take into account the size of the components (the number of users in the component)?

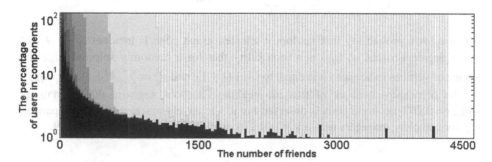

Fig. 2. Percentage of the users in the connected components of a given size for users with a given number of friends.

Figure 2 shows that the total number of users in the largest components considerably exceeds the total number of users in the isolates.

A question arises, how many components of this or that type go to a user? In the network, the following dynamics of the number of the largest connected components is observed: as the number of the friends increases, at first the number of the largest components increases reaching its peak (one largest component per user on the average), and then it decreases up to zero, it is replaced by the next largest component (next size component).

Consequently, we can say that in the user's friends network there is one largest connected component, which includes a greater part of his friends (Fig. 2). Percentage ratios between components of different size can be seen in Fig. 1.

2.4 Friendship of Similar Users

The concept of homophily was introduced into sociology by Paul Lazarsfeld and Robert Merton [1]. It means that if two individuals are similar (they have common characteristics, e.g., age, sex, profession or attitudes), then they are more likely to be connected (e.g., friendship) than if their characteristics differ. The contrary is also true to fact: if two individuals have a link, then they are more likely to possess similar characteristics.

General formulation of the question on homophily is as follows: "What characteristics define homophily in this or that situation?" In case of social network: "What users' characteristics explain the existence of links among them?"

To answer the question if similar users have friendly relations we consider the number of their friends as a characteristic of users' similarity. To make a more precise estimate of the fact that users have friendly relations with similar users there may be used assortativity mixing [2], that takes values from a line segment $[-1, 1]$ (1 – are friends with similar ones, -1 – are friends with dissimilar ones):

$$r = \frac{1}{\sigma_q^2} \sum_{jk} jk(e_{jk} - q_j q_k),$$

where e_{jk} is a probability that randomly selected graph edge is incident to $j + 1$ and $k + 1$ degree vertices, q_j (q_k) is a probability that for a randomly selected edge the degree of vertices incident to it is equal to $j + 1$ ($k + 1$), and $\sigma_q^2 = \sum_k k^2 q_k - (\sum_k k q_k)^2$.

For friendships network of Russian-speaking Facebook segment the assortativity index $r = 0.267$, i.e. homophily is observed but to a relatively low extent. For comparison [2]: for company managers network it is 0.276, while for the Internet it is -0.189.

3 Interdependence of Formal and Informal Relations

In this section we consider interdependence of friendship and commenting in Facebook. We regard commenting link as strong at that if it is proved 9 months a year (i.e. at least one comment each month during 9 months).

We introduce the following event notation: F means that two randomly selected users are friends, C means that there is a link of commenting between two randomly selected users, SF means that two randomly selected users are "strong" friends, SC means that there is a "strong" link of commenting between two randomly selected users.

To define to what extent informal links (commenting links) are conditioned by formal ones (friendship links) we calculate the following probabilities:

- $P(C) = \frac{|E_c|}{|V|*(|V|-1)/2} = 1.3 * 10^{-6}$ and $P(SC) = \frac{|E_{sc}|}{|V|*(|V|-1)/2} = 6.6 * 10^{-9}$ – probabilities that there is a link of commenting between two randomly selected users;

- $P(C|F) = \frac{P(C,F)}{P(F)} = \frac{|E_f \cap E_c|}{|E_f|} = 0{,}041$ and $P(SC|F) = \frac{P(SC,F)}{P(F)} = \frac{|E_f \cap E_{sc}|}{|E_f|} = 0{,}0003$ – probabilities that two users are connected by a commenting relation if there is a friendship link between them;

- $P(C|SF) = \frac{P(C,SF)}{P(SF)} = \frac{|E_{sf} \cap E_c|}{|E_{sf}|} = 0{,}165$ and $P(SC|SF) = \frac{P(SC,SF)}{P(SF)} = \frac{|E_{sf} \cap E_{sc}|}{|E_{sf}|} = 0{,}0013$ – probabilities that two users are connected by a commenting relation if there is a nonrandom (strong) friendship link between them.

The probabilities increase, consequently, friendship links are significant for commenting (strong commenting including). A strong friendship link increases commenting probability by more than 4 times as compared to an "ordinary" friendship link (in particular it indicates the significance of strong friendship links consideration).

To define to what extent formal links (friendship links) are conditioned by informal ones (commenting links) we calculate the following probabilities:

- $P(F) = \frac{|E_f|}{|V|*(|V|-1)/2} = 1.4 * 10^{-5}$ and $P(SF) = \frac{|E_{sf}|}{|V|*(|V|-1)/2} = 3.4 * 10^{-6}$ – probabilities that there is a link of friendship between two randomly selected users;

- $P(F|C) = \frac{P(C,F)}{P(C)} = \frac{|E_f \cap E_c|}{|E_c|} = 0.45$ and $P(SF|C) = \frac{P(SF,C)}{P(C)} = \frac{|E_{sf} \cap E_c|}{|E_c|} = 0.42$ – probabilities that two users are connected by a friendship relation if there is a commenting link between them;

- $P(F|SC) = \frac{P(F,SC)}{P(SC)} = \frac{|E_f \cap E_{sc}|}{|E_{sc}|} = 0.68$ and $P(SF|SC) = \frac{P(SC,SF)}{P(SC)} = \frac{|E_{sf} \cap E_{sc}|}{|E_{sc}|} = 0.67$ – probabilities that two users are connected by a friendship relation if there is a strong commenting link between them.

The probabilities increase, consequently, existence of commenting links indicate friendship (increases its probability). A strong commenting link increases friendship probability by more than 1.5 times as compared to an "ordinary" commenting.

On the whole, the following conclusions can be made:

- commenting links are found considerably more seldom than friendship links;
- if one user comments another one ("ordinary" commenting link), then in half cases they are friends;
- if one user comments another one (strong commenting link), then in two of three cases they are friends;

- if users are friends, the only in one case of 25 there is a commenting link between them;
- if users are friends and they have a common friend, then in one case of six there is a commenting link between them.

Thus, it is not enough to take into account friendship links to study information propagation in the network through commenting.

4 Conclusion

The paper has regarded the network of formal links of Facebook users. General characteristics of the network were defined, a definition to strong friendship links was given, there was analyzed friendship links strength of Facebook users, there was studied the connectivity of user's friends (e.g. in general a bigger part of the user's friends appeared to be in one component, and the other friends turned out to be isolated from other friends), there was found a dependence between the number of friends of the user and of his friends, there was considered interdependence of friendship links and commenting links. The obtained results are planned to be applied in future to modeling of behavior of social network users.

References

1. Lazarsfeld, P.F., Merton, R.K.: Friendship as a social process: a substantive and methodological analysis. In: Berger, M., Abel, T., Page, C.H. (eds.) Freedom and Control in Modern Society, pp. 18–66. Van Nostrand, New York (1954)
2. Newman, M.E.J.: Mixing patterns in networks. Phys. Rev. E **67**(2), 026126 (2003). arXiv: cond-mat/0209450
3. Barnes, J.: Class and committees in a Norwegian Island Parish. Hum. Relat. **7**, 39–58 (1954)
4. Moreno, J.L., Jennings, H.H.: Statistics of social configurations. Sociometry **1**, 342–374 (1938)
5. Granovetter, M.S.: The strength of weak ties. Am. J. Sociol. **78**, 1360–1380 (1973)
6. Granovetter, M.S.: The strength of weak ties: a network theory revisited. In: Marsden, P.V., Lin, N. (eds.) Social Structure and Network Analysis. Sage, Beverley Hills (1973)
7. Kadushin, C.: Understanding Social Networks: Theories, Concepts, and Findings, 252 p. Oxford University Press, Oxford (2012)
8. Wasserman, S.: Social Network Analysis: Methods and Applications, 825 p. Cambridge University Press, Cambridge (1994)

Single-Sentence Readability Prediction
in Russian

Nikolay Karpov$^{(\boxtimes)}$, Julia Baranova, and Fedor Vitugin

National Research University Higher School of Economics,
Nizhny Novgorod, Russia
nkarpov@hse.ru, {ligros7,fedor.vityugin}@gmail.com

Abstract. In an effort to make reading more accessible, an automated readability formula can help students to retrieve appropriate material for their language level. This study attempts to discover and analyze a set of possible features that can be used for single-sentence readability prediction in Russian. We test the influence of syntactic features on predictability of structural complexity. The readability of sentences from SynTagRus corpus was marked up manually and used for evaluation.

Keywords: Natural language processing · Text readability prediction · Single-sentence readability · Syntactic links

1 Introduction

One of integral parts of language teaching is reading, which gives some technical difficulties for professors and students. These difficulties are mainly connected with searching and understanding texts of a concrete level of difficulty (corresponding to a student's knowledge). At the moment there are several research projects that focus on the obtainment of text with the readability level needed for the education purposes. First approach is to classify texts with respect to its level and to retrieve the text needed. Second approach is to take any text and simplify it to the target readability.

This paper describes the part of a project which aim is to develop a system with a simplified functionality. It should be a system of text adaptation to a target level in Russian language as a foreign language (RFL). In the framework of project realization of the automatic simplification of texts in accordance with the language level, we were solving the identification problem of the source and the resulting levels of difficulty of the sentences or texts. Further step will be their lexical and syntactic simplification. In this study we give the results of application of a number of models that identify the level of difficulty of the text or single-sentences using different statistical parameters.

In Sect. 2 **Related work** we presented an overview of the work of the researchers involved in the subject classification of the texts on the basis of complexity of reading in Russian, English and French. In Sect. 3 **Text readability prediction** there are classical models and a model developed specially for the Russian language were adapted to our resources and was tested on texts of several levels of difficulty. Section 4 **Sentence classification** describes the results obtained by applying the formula of Flesch-Kincaid and Dale-Chall to identify the lexical and structural complexity of the

© Springer International Publishing Switzerland 2014
D.I. Ignatov et al. (Eds.): AIST 2014, CCIS 436, pp. 91–100, 2014.
DOI: 10.1007/978-3-319-12580-0_9

Russian sentences. In Sect. 5, **Sentence classification with syntactic features** we give one of the variants of the model for the effective identification of readability of Russian sentences with the use of syntactic features. In Sect. 6 **Conclusion** there are general conclusions on the executed experiments and plans for further work and improvement of the models considered.

2 Related Work

The first studies on text complexity started in the 20's of the past century. This field of research was mainly developed in the field of work relating to the English language, but over the last decade a number of works related to other languages were worried out, that testifies to the fact, that the research relating to automatic identification of the complexity of the text is still highly relevant.

The complexity of reading can be represented as a function that finds a correspondence to a certain level of complexity of the predefined text with a variety of variables, extracted from the text. Traditionally variables allocated for the characterization of these texts are divided into two groups – lexical parameters and syntactic parameters. In one of the most common formula – the formula of Flesch-Kincaid [1, 2] the complexity of the text is represented as a linear function of the average number of syllables per word and the average length of the sentence.

A formula of Dale and Hall [3] also defines a syntax difficulty of the text as the average length of the sentence, but for lexical metric it uses the percentage of words not from list of 3000 Easy Words, which is based on familiarity of words. This means that all the words in the list are familiar to US children in the 4th grade.

With the growth of computing power there appeared an opportunity to build more complex models. Model of Collins-Thompson and Callan (2005) [4] uses frequency of words unigrams (a dictionary is specified for each level of the language) and features that some words are most possible for prediction a certain level of complexity of the text. Schwarm and Ostendorf [5] use more complex syntax parameters - the average height of trees parsing, the number of nominal and verbal groups, the average number of non terminal nodes and so on.

Automatic identification of a reading difficulty in Russian language is also researched in a number of works. Oborneva (2006) in her work [6] adapts the formula of Flesch and Flesch-Kincaid for the Russian language by means of adjustment coefficients: she compares the average length of syllables in English and Russian words and percentage of multi-syllable words in dictionaries for these languages. It is also worth noting the study of Krioni, Nickin and Philippova who define the complexity of educational texts in Russian language highlighting a number of more complex parameters of assessed texts: connectivity, structure, integrity, functional and semantic type, information, abstractness of the text presentation and complexity of linguistic structures [7].

Due to the large amount of research dedicated to readability assessment, we have highlighted only the most eminent works. Nevertheless all of them identify the difficulty of reading of the whole text. Our goal is to determine the efficiency of the developed techniques in relation to the texts in general and sentences in particular in

Russian language as well as checking our own developed model to determine the difficulty of sentence reading.

3 Text Readability Prediction

First task was to perform the prototyping of Russian text retrieval with needed readability. The main goal of this process was to find which kind of variables and classification algorithm would allow us to obtain the highest indicators of precision and recall of readability prediction. There was conducted a series of experiments on the training of different classification algorithms. We experimented with the following algorithms:

- naive Bayes;
- k-nearest neighbors;
- classification tree;
- random forests;
- SVM.

Evaluation was performed with the help of cross validation on the test part of our collection. We extract features from a collection consists of 219 texts divided into four groups. Levels distribution is following: A1 (elementary – 52), A2 (basic) – 57, B1 (first) – 60, C2 (difficult) – 50 according to levels described in Common European Framework of Reference for Languages (CEFR) [8]. The first three groups include texts, created specially for second language learners of Russian, with respect to their level of language knowledge on the basis of news articles.[1] Fourth group (difficult) consists of original news for native readers. We extract 25 variables from texts proposed in the previous works:

1. Average number of words in the sentence of the text.
2. Average length of one word in a sentence.
3. Text length in letters.
4. Text length in words.
5. Average sentence length in syllables.
6. Average length of words in syllables.
7. Percentage of words with number of syllables more or equal to N. We define N as each value from 3 to 6.
8. Average sentence length in letters.
9. Average length of words in letters.
10. Percentage of words with number of letters more or equal to N. We define N as each value from 5 to 13.
11. The percentage of words in a sentence, not included in the active vocabulary of A1 level.
12. The percentage of words in a sentence, not included in the active vocabulary of A2 level.

---

[1] http://texts.cie.ru

13. The percentage of words in a sentence, not included in the active vocabulary of B1 level.
14. The occurrence in the sentence of concrete parts of speech.

We mark seventeen parts of speech in the texts according to the list of grams in the OpenCorpora [9]: noun (NOUN), full form of an adjective (ADJF), short form of an adjective (ADJS), comparative (COMP), personal form of the verb (VERB), infinitive form of the verb (INFN), full participle (PRTF), short participle (PRTS), gerund (GRND), numeral (NUMR), adverb (ADVB), noun-pronoun (NPRO), predicative (PRED), preposition (PREP), conjunction (CONJ), a particle (PRCL), interjection (INTJ). We were interested in occurrence of parts of speech as proposed by Francois, 2009 [10].

We did not use some variables described in paper [11] due to adaptation to our texts. We did not use variable connected with paragraph because our texts are very short. Texts do not have syntactic markup that is why the concept of a phrase was not used either.

First experiment was a binary classification of readability: A1 versus C2, A2 versus C2, B1 versus C2. With the help of Classification Tree, SVM and Logistic Regression algorithms the accuracy we got was really high, it was almost equal to 1.

Second experiment for texts classification of four levels got lower accuracy. An example of accuracy of text retrieval with B1 level of readability is shown in Table 1.

Table 1. Results of texts retrieval with B1 readability level.

Method	Classification accuracy	F-measure	Precision	Recall
SVM	0.8092	0.7965	0.8491	0.75
Classification Tree	0.9905	0.9916	1	0.9833
kNN	0.8131	0.7333	0.7333	0.7333
Random Forest	0.9818	0.9667	0.9667	0.9667
Naive Bayes	0.8726	0.7890	0.8776	0.7167

kNN is a K nearest neighborhood method. Results received during the second experiment are worse than the first experiment with only two levels. Due to the fact that results of the Classification Tree method reached 99 %, we can say that the obtained results meet the needs.

To analyze the effect of each variable for the texts discrimination into 4 levels we ranked it by information gain ratio [12] (Table 2).

The first three variables have the highest information gain ratio; they are lexical ones. We can say that they are most important variables for discrimination.

4 Sentence Classification

Next task was to make a prototype of an algorithm to retrieve difficult sentences for further simplification. This algorithm is based on a sentence classification with respect to its readability. For results evaluation we use subcorpus of Russian national Corpus

Table 2. Texts variables ranked by information gain ratio (top 10).

Variable name	Information gain ratio
The percentage of words in a sentence, are not included in the active vocabulary of A1 level	0.105141
The percentage of words in a sentence, are not included in the active vocabulary of A2 level	0.105141
The percentage of words in a sentence, are not included in the active vocabulary of B1 level	0.084211
Percentage of words with 8 letters or more	0.040098
Percentage of words with 9 letters or more	0.038431
Percentage of words with 7 letters or more	0.036923
Average sentence length in syllables	0.034359
The average length of one word in a text	0.034359
Percentage of words with 10 letters or more	0.033689
Percentage of words with 5 syllable and more	0.033193

(RNC) – corpus SunTagRus [13] that has morphological and syntactic metadata. We manually tagged 3500 sentences from this subcorpus to mark their structural level of perception complexity. We found out that level B1 suits the majority of our students. So, we created a binary sentence markup, which is (1) B1 or lower than B1 and (2) Higher than B1.

Lexical difficulty markup was made on the basis of active vocabulary of three levels: A1, A2 and B1. The most complete vocabulary list (B1) includes 2500 words. So, we defined sentences having more than 33 % active vocabulary words as lexically difficult ones.

Thus, we have two kinds of markup: structural complexity and lexical difficulty. As an intersection of its lexical and structural level of difficulty we obtained markup of a total level of difficulty.

Dale-Chall model was developed to define the difficulty of text with the help of linear function of flowing variables: average sentence length (number of words divided by number of sentences) and rear words in the text.

When we use these variables for single sentence readability prediction we need to adapt them as following: sentence length rather than average sentence length, percentage of words not in the active vocabulary with respect to sentence length (number of words in the sentence not in the vocabulary divided by total number of words in the sentence) instead of rear words percentage. In our case, we don't need to use dictionary of the Russian language with the frequency of words occurrence because we have a definite list of words that are contained in active vocabulary.

These two variables were automatically extracted for each sentence in our corpus. We predict readability for single sentence using different methods of machine learning as shown in Sect. 3. Evaluation was performed with the help of cross validation on the test part of corpus.

To evaluate the influence of each variables first we try to predict difficulty by using variables separately, next we predict it using both variables. It is easy to see that even in the case of prediction with the help of sentence length we can obtain good results. But if we need to classify to more than two numbers of levels, accuracy will decrease. Precision of difficult sentence retrieval is lower than simple sentence retrieval.

Accuracy of readability prediction on the basis of both variables is much higher. The second variable - percentage of words not in the active vocabulary cut off many difficult short sentences. It is effect to the precision of difficult sentence retrieval. The results are presented in Table 3.

Table 3. Results of readability prediction using variables: sentence length and percentage of words not in the active vocabulary.

Method	Classification accuracy	F-measure (difficult/ simple)	Precision	Recall
Naive Bayes	0.8846	0.9242/0.7581	0.9378/0.7246	0.9110/0.7950
Logistic regression	0.8745	0.9212/0.6921	0.8945/0.7833	0.9495/0.6199
kNN	0.8941	0.9299/0.7840	0.9519/0.7318	0.9089/0.8441
Random Forest	0.8840	0.9208/0.7837	0.9747/0.6808	0.8725/0.9233
Classification Tree	0.8955	0.9308/0.7866	0.9527/0.7347	0.9099/0.8465

We have opposite situation in this case. Precision of difficult sentence retrieval is higher than simple sentence retrieval. We can conclude that even using only these two variables we can effectively predict sentence readability.

Flesch-Kincaid model grade level formula was also used to determine readability. The formula utilized in the software is $[(0.39 \times ASL) + (11.8 \times ASW) - 15.59]$, where ASL is the average sentence length (number of words divided by number of sentences) and ASW is the average syllables per word (number of syllables divided by number of words). To apply this formula to the problem of estimating the difficulty of the single sentence we can save ASW in its original form and instead of ASL use sentence length (number of words).

If we come to analyze how the lexical difficulty is predicted with the help of the average syllables per word (ASW) it is easy to notice that ASW exert to classification accuracy of difficult sentences and not exert to simple one. The reason is that Russian language is characterised by the presence of many long words (with many syllables), which are simple ones because they are created by combining short words. This is the help of two variables (ASL and ASW) we get results that are shown in Table 4.

Total accuracy for only two variables is relatively high but the recall of simple sentences retrieval is quite low. Active vocabulary in the first certified level of Russian language could not be exactly determined using the average syllables per word.

Table 4. Results of difficult/simple sentence retrieval from text using ASL and ASW.

Method	Classification accuracy	F-measure (difficult/simple)	Precision (difficult/ simple)	Recall (difficult/ simple)
Naive Bayes	0.7967	0.8794/0.3550	0.8119/0.6386	0.9590/0.2458
Logistic regression	0.7945	0.8770/0.3761	0.8156/0.6086	0.9484/0.2722
kNN	0.7746	0.8640/0.3434	0.8093/0.5094	0.9265/0.2590
Random Forest	0.7910	0.8788/0.2431	0.7961/0.6910	0.9806/0.1475
Classification Tree	0.7801	0.8669/0.3673	0.8140/0.5318	0.9272/0.2806

5 Sentence Classification Using Syntactic Structure

We use deeper sentence features, which potentially can improve accuracy of readability prediction – syntactic relations of words. Our experiment was carried out on the basis of SynTagRus corpus, which has morphological and syntactic metadata. We decide to use syntactical features of a sentence as a basis of classification algorithm because this approach shows better results on the preliminary stage whether morphology features or n-gramms. In this case on the basis of syntactical features classification tasks look as follows. The sentences are tagged with morphological metadata using OpenCorpora [9]. On the basis of morphological marks we generate syntactical links. Its syntactical links help us to predict single sentence readability.

SynTagRus includes about 60 types of syntactic links grouped as it proposed in RNC. We try to predict sentence readability with the help of two data representation. First we use all 60 types of syntactic links. We get following experimental results shown in Table 5.

Table 5. Classification using 60 types of links.

Method	Classification accuracy	F-measure	Precision	Recall
Naive Bayes	0.7570	0.7459	0.7813	0.7136
Logistic regression	0.7112	0.7077	0.7160	0.6995
kNN	0.7286	0.7146	0.7531	0.6798
Random Forest	0.7582	0.7472	0.7822	0.7153
Classification Tree	0.7047	0.6414	0.8158	0.5284

Then we use aggregated links to 4 groups as it proposed in RNC. Classification accuracy using aggregated variables was lower. On the basis of obtained experimental results it was concluded that we should use all types of links without aggregation. The best precision and recall showed SVM algorithm.

It is obvious to assume that syntactic variables can predict structural difficulties better. Thus we used the same approach as it was with other previous models, perform experiment with structural and lexical difficulty separately. Results are presented in Table 6.

Table 6. Results of structural difficulties prediction using only syntactic variables.

Method	Classification accuracy	F-measure (difficult/ simple)	Precision	Recall
Naive Bayes	0.8085	0.8021/0.8144	0.8244/0.7942	0.7810/0.8356
kNN	0.7681	0.7128/0.8055	0.9271/0.6965	0.5790/0.9550
Classification Tree	0.8180	0.8056/0.8289	0.8589/0.7860	0.7585/0.8768
SVM	0.7956	0.8010/0.7900	0.8972/0.8173	0.9174/0.7645
Random Forest	0.8374	0.8307/0.8436	0.8610/0.8170	0.8271/0.8719

We can conclude that syntactic variables allow predicting structural difficulties more efficiently than simple variables. Next we use all kind of variables (syntactic and lexical) to predict total difficulty of sentence. As a lexical variable we use percentage of words not from active vocabulary of the corresponding level (Table 7).

Table 7. Results of total readability prediction using all kinds of variables and syntactic links.

Method	Classification accuracy	F-measure (difficult/ simple)	Precision	Recall
Naive Bayes	0.8191	0.8906/0.4767	0.8354/0.6975	0.9537/0.3621
kNN	0.8224	0.8893/0.5501	0.8571/0.6493	0.9241/0.4772
Random Forest	0.9443	0.9640/0.8768	0.9620/0.8832	0.9661/0.8705
Classification Tree	0.9364	0.9584/0.8648	0.9679/0.8380	0.9491/0.8933
SVM	0.8633	0.9125/0.6875	0.9679/0.7165	0.9491/0.6607

Last approach gives more stable results and may be used to increase the number of classes of sentence complexity (Table 8).

Table 8. Results of total readability prediction using all kinds of variables and syntactic links.

Variable name	Information gain ratio
The percentage of words in a sentence, are not included in the active vocabulary of B1 level	0.318
Sentence length in letters	0.122
Percentage of words with 3 syllable and more	0.119
Sentence length in syllables	0.118
Sentence length in words	0.098
Syntactic predicative link	0.095
Average words length in syllables	0.092
The average length of one word in a text	0.092
Percentage of words with 7 letters or more	0.069
Percentage of words with 5 letters or more	0.069

6   Conclusion

Classical models and models developed specially for Russian language were adapted to news texts retrieval. These models give good results. We managed to develop a precise classification system of news texts in Russian with respect to their readability.

Accuracy of four levels classification was lower. Due to the fact that obtained results of the Classification Tree and Random Forest methods reached 99–98 %, we can say that they met our needs.

We adapted traditional classification techniques with statistical features like Flesch-Kincaid and Dale-Chall to identify lexical and structural complexity of Russian sentences. These techniques were tested on set of sentences where readability was manually marked as binary classification.

Finally, we found one of the variants of the model for the effective identification of readability of Russian sentences with the use of syntactic links. We found that syntactic features can predict structural complexity. Total set of features with statistical, lexical and syntactical ones can predict sentence readability with 0.9661 amount of recall using Random Forest algorithm. Most important features for this classification are lexical ones.

Acknowledgment. This study comprises research findings from the «Adaptation of texts from the Russian National Corpus» for the electronic textbook «Russian language as a foreign one» carried out within The National Research University Higher School of Economics' Academic Fund Program in 2013, grant No 13-05-0031.

References

1. Flesch, R.: A new readability yardstick. J. Appl. Psychol. **32**, 221 (1948)
2. Kincaid, J.P., Fishburne, Jr., R.P., Rogers, R.L., Chissom, B.S.: Derivation of new readability formulas (automated readability index, fog count and flesch reading ease formula) for navy enlisted personnel. DTIC Document (1975)
3. Chall, J.S.: Readability Revisited: The New Dale-Chall Readability Formula. Brookline Books Cambridge, MA (1995)
4. Collins-Thompson, K., Callan, J.: Predicting reading difficulty with statistical language models. J. Am. Soc. Inf. Sci. Technol. **56**, 1448–1462 (2005)
5. Schwarm, S.E., Ostendorf, M.: Reading level assessment using support vector machines and statistical language models. In: Proceedings of the 43rd Annual Meeting on Association for Computational Linguistics, pp. 523–530. Association for Computational Linguistics (2005)
6. Oborneva, I.: Automatic assessment of the complexity of educational texts on the basis of statistical parameters (2006)
7. Krioni, N., Nikin, A., Filippova, A.: Automated system for analysis of the complexity of educational texts. Manag. Soc. Econ. Syst. **11**, 101–107 (2008)
8. Verhelst, N., Van Avermaet, P., Takala, S., Figueras, N., North, B.: Common European Framework of Reference for Languages: Learning, Teaching. Assessment. Cambridge University Press, Cambridge (2009)
9. Bocharov, V., Stepanova, M., Ostapuk, N., Bichineva, S., Granovsky, D.: Quality assurance tools in the OpenCorpora project. In: Computational Linguistics and Intelligent Technology: Proceeding of the International Conference « Dialog–2011 », pp. 10–17 (2011)

10. Francois, T.L.: Combining a statistical language model with logistic regression to predict the lexical and syntactic difficulty of texts for FFL. In: Proceedings of the 12th Conference of the European Chapter of the Association for Computational Linguistics: Student Research Workshop, pp. 19–27. Association for Computational Linguistics (2009)
11. Nevdah, M.: Development of a method of automated evaluation of the complexity of educational texts for higher school (2008)
12. Kent, J.T.: Information gain and a general measure of correlation. Biometrika **70**, 163–173 (1983)
13. Nivre, J., Boguslavsky, I.M., Iomdin, L.L.: Parsing the SynTagRus treebank of Russian. In: Proceedings of the 22nd International Conference on Computational Linguistics, vol. 1, pp. 641–648. Association for Computational Linguistics (2008)

Matchings and Decision Trees
for Determining Optimal Therapy

Natalia Korepanova[1](✉), Sergei O. Kuznetsov[1],
and Alexander I. Karachunskiy[2]

[1] School of Applied Mathematics and Information Science,
National Research University Higher School of Economics, Moscow, Russia
korepanova.natalia@gmail.com, skuznetsov@hse.ru
[2] Research and Clinical Center of Pediatric Hematology,
Oncology and Immunology, Moscow, Russia
aikarat@mail.ru

Abstract. An approach to the study of different types of treatments in subgroups is proposed. This approach is based on matching algorithms and decision trees. An application to the data on children with acute lymphoblastic leukaemia is considered.

Keywords: Medical informatics · Decision trees · Optimal therapy · Machine learning for medicine

1 Introduction

Nowadays one of the most promising way of therapy optimization, especially in pediatric haematology, is conforming a therapy to various subgroups of patients which are described by patients' physiological features. Usually, the number of possible subgroup descriptions is large, and often physicians are able to chose subgroups for analysis relying just on their experience and observations. Therefore, in statistical terms any subgroup analysis which is aimed at showing the superiority in efficiency of one treatment strategy over one or several others seems doubtful as a rule. In the present paper the approach of finding subgroups with significantly different or equivalent response to two treatment strategies is proposed. Obtained hypotheses can underlie subgroup analysis with better choice of subgroups.

The analysis is carried out for the database on children with acute lymphoblastic leukemia (ALL) [1] who underwent a course of one of two types of induction therapy. The first step consists in finding the largest set of pairs of similar patients who took different drugs with the help of the Gale-Shapley algorithm [2-4] for computing optimal stable matching. This algorithm is based on the concept of physiological "similarity" between two patients; therefore, the definition of "distance" between two patients is introduced. After that the derived matching is examined for the existence of the classes in which treatment strategy strongly affects or does not affect treatment response. At the second step

© Springer International Publishing Switzerland 2014
D.I. Ignatov et al. (Eds.): AIST 2014, CCIS 436, pp. 101–110, 2014.
DOI: 10.1007/978-3-319-12580-0_10

we attempt to describe extracted classes by applying decision trees with various parameters [5,6]. It appears that decision trees are not able to describe every class on the whole. However, some subgroups of patients for whom one drug is more appropriate than another one can be selected from the results of classification. Moreover, decision trees allow us to formulate hypotheses about comparison of therapy efficiencies in subgroups in form suitable for haematologists. And finally, results are approved or disapproved by classical medical statistical methods.

The rest of the paper is organized as follows. In Sect. 2 the dataset is described. In Sect. 3 the steps of proposed approach are presented in detail. In Sect. 4 the application to the initial dataset and its results are shown. Section 5 concludes the paper.

2   Dataset

The dataset consists of 1946 patients up to 19 years old of age with newly diagnosed acute lymphoblastic leukaemia (ALL). This dataset is stored as a database containing the following data fields: sex (male or female), age (in years), initial white blood count (per nl) (WBC), immuno-phaenotype (8 types), CNS status (3 types), palpable liver size (in cm), palpable spleen size (in cm), mediastinum status (3 types), date of allocation to treatment, last status report (alive, no information, death), date of the latest follow up visit, treatment strategy (2 types).

The analysis was based on the comparison of the efficiency of two treatment strategies: under DEXA $6\,mg/m^2/d$ and MePRED $60\,mg/m^2/d$ [1], which we call S1 and S2. To find relations between initial characteristics and survival rate all physiological features presented above were chosen: sex, age, initial WBC, immuno-phaenotype, CNS, palpable liver size, palpable spleen size, mediastinum status. To evaluate the therapy efficiency overall survival [7] was calculated with death as the event. Survival time was calculated from diagnosing until the date of last status report. If the value of one patient's characteristic was not determined, this patient was not included in analysis. Consequently, we obtained data on 1535 appropriate patients: 939 of them were assigned S1, and 596 of them were assigned S2.

3   Proposed Approach

Our procedure is intended to find subgroups where differences between two competing treatment strategies are noticeable or do not exist. The input data consist of two sets of patients corresponding to the strategies. All patients are described by several initial features each could be either numerical or categorical. The number of features can be reduced applying selection feature techniques, or relying on expert views. In the current research we were provided with information on 8 features which are the most influential in haematologists' sight, therefore feature selection techniques were not required. Nevertheless it can surprisingly appear that unprovided features affect intensively the survival time. So, as for medical data, we believe that results of feature selection and expert views should be combined. Let us now move on to the steps of the proposed procedure.

3.1 Patient Distance

First of all, the procedure suggests defining distance between two patients (the inverse concept to "physiological similarity"). If all physiological features are numerical, it is possible to use one of the classical distance measures [8–10]. However, there are likely several categorical features in patient descriptions. Therefore it is required to modify classical definitions of the distance. It is considered that there is no sense to measure distance between two patients that cannot be compared. For this purpose it is necessary to give a definition of "comparability" of two patients.

Definition 1. *Two patients are **comparable** if the values of all their categorical physiological features coincide. If they differ in just one categorical feature, they are **incomparable**. So, if all initial features of patients are numerical, any two patients are comparable.*

Before computing distance between comparable patients, numerical features need to be normalized to make all feature impacts equivalent. Therefore all of them are centered by subtracting the mean value and then scaled by dividing by the range of values [11]. So, the distance between two comparable patients is computed based on the normalized values of their numerical features.

3.2 Pairs of Similar Patients

The distances between all pairs of patients where the first one is from the first set and the other one is from the second set are computed. To find pairs of similar patients *the deferred acceptance procedure* [2–4] is applied for two sets of patients who underwent different courses of treatment. This algorithm was developed to solve the marriage problem, i.e. the problem of finding stable matching. It is applicable to two sets of instances which are often referred to as men and women. Every man ranks women and every woman ranks men in accordance with their preferences. Thereupon each man proposes to his favourite woman, and each woman rejects all but her favourite, who becomes her marriage nominee. The rejected men propose to their next choices, and each woman chooses her favourite among the new proposers and the nominee rejecting all the rest, and so on. As soon as no men are rejected or they have no more choices each woman accepts her nominee. Eventually, we get the pairs consisting of one man and one woman. In other words, every pair includes two instances from different sets. The result of the algorithm application is stable, and optimal if preferences are complete. In case of one-to-one matching completeness also accounts for uniqueness of the provided matching. Before applying this algorithm to medical data definitions of preference and completeness should be given.

Definition 2. *Patient p* **prefers** *patient q_1 to patient q_2 if the distance between p and q_1 is less than the distance between p and q_2. Patient p is* **indifferent** *between patients q_1 and q_2 if the distances between p and q_1 and between p and q_2 are equal.*

Definition 3. *Preferences are* **complete** *if for every x, y from one set and for every z from the other one, z prefers x to y, or y to x, or it is indifferent between them.*

Some patients may be incomparable w.r.t preference. However, it results from the definition of comparability that all patients can be partitioned into subsets where all patients are comparable with each other. Consequently, in these subsets patients preferences are complete, and the result of *the deferred acceptance procedure* application to every such subset is unique, stable and optimal [4], which means that the constructed matching is unique, stable and optimal in total.

3.3   Separation into Classes in Terms of Efficiency (Overall Survival)

Using the matching and patients? survival times we can determine classes of patients with quite clear or without any dissimilarities in survival time under treatment strategies. The simplest approach is to visualize the matching in any way and attempt to mark boundaries of such classes. Therefore, it is proposed to consider the coordinate plane where X-axis is survival time under the first curing strategy and Y-axis is survival time under the second one. Every pair in the matching is associated with a point on the plane. The first coordinate of the point is the survival time of the patient who has received the first kind of treatment, and the second one is the survival time of the patient who has received the other one.

All points (pairs of patients) are partitioned into several classes according to sensitivity to treatment strategies (e.g., survival time under treatment strategy 1 is superior to that under treatment strategy 2, survival time under treatment strategy 2 is superior to treatment strategy 1, short survival time under both treatment strategies, and long survival time under both of them). Further manipulations are carried out on data about individual patients.

3.4   Hypotheses Generation and Verification

It is insufficient to separate patients into classes in which survival times under different strategies differ or do not differ. It is more essential to obtain descriptions of these classes, so the classification problem arises. In the case of comparing treatment strategy efficiencies decision trees with various parameters [5,6] seem appropriate because, in general, the accuracy of the other well-known methods is lower on the initial data. Moreover, the form of hypotheses generated by decision trees is comprehensible for physicians. So, in our computer experiments we used information gain, information gain ratio and Gini index as attribute selection criteria [12,13]. Also, minimum number of instances in leaves, maximal allowable tree depth and sufficient percent of majority class for nonspliting were varied. The approach evaluation was conducted by means of 10-fold cross-validation.

Decision trees output data are descriptions of classes in terms of characteristics of the patients belonging to these classes. Those descriptions may be transformed into hypotheses about the existence or the absence of the difference in treatment strategy efficiencies. To show how it works, assume that any description of the class with the superiority of the first treatment strategy has been received. This assumption can be transformed into the following hypothesis: for the patients who fall under obtained description overall survival under the first strategy is higher than that under the second one. This sort of hypotheses are put forward on the basis of the most evident subgroups output by decision trees.

All formed hypotheses are tested by classical medical statistical tools. The first of them is Kaplan-Meier survival curves [14–16] which estimate sample survival rate functions for censored data. The second one is log-rank test [16,17], a nonparametric hypothesis test to compare the survival distribution of two samples with no-difference null hypothesis and standard normally distributed statistics. In contrast to log-rank, the equivalence test [18–20] is applied to confirm that survival rates of two samples do not differ, and usually used if log-rank null hypothesis has not been rejected. Also for each hypothesis false negative error (type II error) is computed [21]. It is important to mention that hypotheses are tested on the set of patients that consists not only of those who have been included in classification, but also of those who have not been matched with anybody or have not been labeled with any class mark. If a hypothesis is confirmed by the tests and false negative error is not very large, then it can be analysed by physicians in further random trials. The necessity of new trials is specified by the worldwide statistical principles of clinical trials. According to the notes of European Medicines Agency [22], any clinical trials may have two aspects: confirmatory and exploratory ones. For the first of them the hypotheses are pre-defined, and are tested when the trial is complete, while the second aspect allows of the data dependent choice of hypotheses, and the ability of changes in response to accumulating results. Obviously, the proposed procedure is intended for the exploratory aspect of trials, therefore its results "cannot be the basis of the formal proof of efficacy" [22]. However, the exploratory investigations can serve "for suggesting further hypotheses for later research" [22]. The last statement clearly explains the main purpose of the proposed approach.

4   Analysis and Results

According to the proposed approach it is necessary to distinguish between numerical and categorical physiological features. The initial dataset contains four numerical features: age, initial WBC, palpable liver size, palpable spleen size; and four categorical features: sex, immuno-phaenotype, CNS, mediastinum status. The second ones determine the comparability of patients. However, immuno-phaenotype has two levels of categorization: B- or T-ALL (nominal values), and each of these types has four ordinal subcategories. Therefore, the condition of

Table 1. Numerical values of immuno-phaenotype.

B	−1
pre-B	−0.75
common-B	−0.5
pre-pre-B	−0.25
early-T	0.25
intermediate-T	0.5
mature-T	0.75
hybrid	1

comparability was weakened for this feature. Thus, two patients were comparable if both of them had B- or T-ALL and values of all other categorical features coincided.

To take into account the second level of immuno-phaenotype categorization, its values were transformed into numerical values according to Table 1.

So, for child-ALL data distance was computed using normalized values of numerical characteristics and numerical values of immuno-phaenotype. As it is mentioned before, any standard definition of distance can be chosen. Therefore, Manhattan [8,9], Euclidean [8,9], Minkowski [8,10] with factor 3, Minkowski with factor 100 and Chebyshev [8] distances were used in computer experiments. We found out that there was no meaningful difference between these measures, so for further analysis and method specification Euclidean distance was used.

After that the deferred acceptance procedure was applied, the scatter plot of the derived matching is shown in Fig. 1. There were about 5 most evident isolated classes of points (Fig. 2). As for classes 1 and 2 survival time under both strategies was not long. However, it was slightly longer under S1 in class 1 and slightly longer under S2 in class 2. We can also say that for class 3 the survival time under S1 was longer than under S2 and vice versa for class 4. The survival time for class 5 was long under both strategies. We attempted formalizing boundaries in the way shown in Fig. 3. It is important to mention that the 4-years boundary was not selected randomly. There are about 4 years between the latest diagnosing date and the latest last status report date. In other words, if survival time of any patient was shorter than 4 years this patient was certainly dead or escaped from the observation. For the sake of clear separation of the classes the points between dashed lines were excluded from the analysis.

By applying decision trees to all presented partitions we obtained several hypotheses, one of the most reliable hypothesis is presented below.

The hypothesis is *"MePRED is more efficient than DEXA for patients who are equal to or more than 6.6 years old, with palpable spleen size not smaller than 3.5 cm, and pre-pre- or pre-B immuno-phaenotype"*. There were 39 patients of that kind for classification and 47 such patients at all. This subgroup is not numerous, but, at first, Kaplan-Meier curves (Fig. 4) seem to confirm the hypothesis. The value of log-rang statistics is equal to 2.12 which allows one to reject

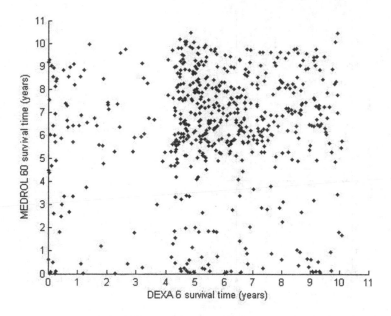

Fig. 1. Scatter plot of all matching pairs.

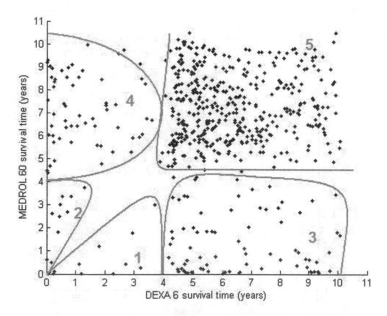

Fig. 2. Scatter plot of all matching pairs with outlined classes.

the hypothesis about no difference at confidence level of 0.95. The false negative
error amounts to 0.31. This is quite good, so, we can propose this hypothesis to
test in further clinical random confirmatory trials.

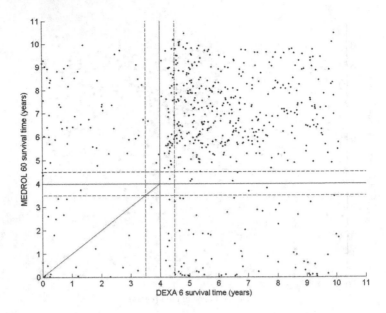

Fig. 3. Scatter plot of all pairs with partition into 5 classes.

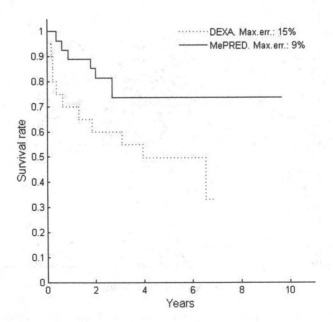

Fig. 4. Kaplan-Meier curves for the case of 6.6 years old and older patients with palpable spleen size not smaller than 3.5 cm and pre-pre- or pre-B immuno-phaenotype.

5   Conclusion

In this paper we introduced a novel approach to solving the problem of determining relevant subgroups of patients for therapy optimization. Getting the dataset of patients described by their physiological characteristics, dates of diagnosing and last status report, the procedure constructs the optimal stable matching between patients who took different drugs and attempted to describe subgroups in which the efficiency of these drugs are different or approximately equal. In further studies other learning techniques will be used, in particular, those based on closed descriptions [23–25].

The proposed procedure can also be applied in other studies of subgroup analysis. Moreover, all parts of the procedure are flexible to changes and can be adapted to other practical problems of subgroup analysis. The main idea of this work consists in proposing the order in which data analysis techniques can be applied, and how they can influence any therapy optimization. Hopefully, the obtained hypotheses will be successfully used in Russian ALL-treatment studies.

References

1. Karachunskiy, A., Herold, R., von Stackelberg, A., et al.: Results of the first randomized multicenter trial on childhood acute lymphoblastic leukaemia in Russia. Leukemia **22**, 1144–1153 (2008)
2. Gale, D., Shapley, L.S.: College Admissions and the Stability of Marriage. Am. Math. Mon. **69**(1), 9–15 (1962)
3. Roth, A.E.: Differed acceptance algorithm: history, theory, practice, and open questions. Int. J. Game Theory **36**(3–4), 537–569 (2007)
4. Alkan, A., Gale, D.: Stable schedule matching under revealed preference. J. Econ. Theory **112**, 289–306 (2003)
5. Fürnkranz, J.: Decision tree. In: Sammut, C., Webb, G.I. (eds.) Encyclopedia of Machine Learning, pp. 263–267. Springer, New York (2010)
6. Rokach, L., Maimon, O.: Classification trees. In: Maimon, O., Rokach, L. (eds.) Data Mining and Knowledge Discovery Handbook, 2nd edn, pp. 149–174. Springer, New York (2010)
7. NCI Dictionary of Cancer Terms. http://www.cancer.gov/dictionary?cdrid=655245. Accessed 7 March 2014
8. Deza, M.M., Deza, E.: Encyclopedia of Distances, pp. 94, 323–324. Springer, Heidelberg (2009)
9. Shekhar, S., Xiong, H.: Distance measures. In: Encyclopedia of GIS, p. 245. Springer, New York (2008)
10. Fuhrt. B.: Distance and similarity measures. In: Encyclopedia of Multimedia, pp. 188–189. Springer, New York (2008)
11. Mirkin, B.G.: Core Concepts in Data Analysis: Summarization, Correlation, Visualization. Springer, London (2011)
12. Raileanu, L.E., Stoffel, K.: Theoretical comparison between the Gini Index and Information Gain criteria. Ann. Math. Artif. Intell. **41**, 77–93 (2004)
13. Kotsiantis, S.B.: Decision trees: a recent overview. Artif. Intell. Rev. **39**, 261–283 (2013)

14. Kaplan, E.L., Meier, P.: Nonparametric estimation from incomplete observations. J. Am. Stat. Assoc. **53**(282), 457–481 (1958)
15. May, W.L.: Kaplan-Meier survival analysis. In: Schwab, M. (ed.) Encyclopedia of Cancer, pp. 1590–1593. Springer, Heidelberg (2009)
16. Kleinbaum, D.G., Klein, M.: Kaplan-Meier survival curves and the log-rank test. In: Kleinbaum, D.G., Klein, M. (eds.) Survival Analysis, pp. 55–96. Springer, New York (2012)
17. Beyersmann, J., Schumacher, M., Allognol, A.: Nonparametric hypothesis testing. In: Competing Risks and Multistate Models with R, pp. 155–158. Springer, New York (2012)
18. Piaggio, G., Elbourne, D.R., Altman, D.G., Pocock, S.J., Evans, S.J.W., for the CONSORT Group: Reporting of noninferiority and equivalence randomized trials: extension of the CONSORT 2010 statement. JAMA **308**(24), 2594–2604 (2012)
19. Machin, D., Gardner, M.J.: Calculating confidence intervals for survival time analyses. Brit. Med. J. **296**, 1369–1371 (1988)
20. Goberg-Maitland, M., Frison, L., Halperin, J.L.: Active-control clinical trials to establish equivalence or noninferiority: methodological and statistical concepts linked to quality. Am. Heart J. **146**(3), 398–403 (2003)
21. Glanz, S.A.: Primer of Biostatistics, 7th edn. McGraw-Hill Education, New York (2011)
22. ICH Topic E9: Statistical Principles for Clinical Trials. Step 5. (2.1 Trial Context.), pp. 6–7 http://www.ema.europa.eu/docs/en_GB/document_library/Scientific_guideline/2009/09/WC500002928.pdf. Accessed 7 March 2014
23. Ganter, B., Kuznetsov, S.O.: Hypotheses and version spaces. In: Ganter, B., de Moor, A., Lex, W. (eds.) ICCS 2003. LNCS (LNAI), vol. 2746. Springer, Heidelberg (2003)
24. Blinova, V.G., Dobrynin, D.A., Finn, V.K., Kuznetsov, S.O., Pankratova, E.S.: Toxicology analysis by means of the JSM-method. Bioinformatics **19**(10), 1201–1207 (2003)
25. Ganter, B., Grigoriev, P.A., Kuznetsov, S.O., Samokhin, M.V.: Concept-based data mining with scaled labeled graphs. In: Wolff, K.E., Pfeiffer, H.D., Delugach, H.S. (eds.) ICCS 2004. LNCS (LNAI), vol. 3127, pp. 94–108. Springer, Heidelberg (2004)

Indicators of Connectivity for Urban Scientific Communities in Russian Cities

Fedor Krasnov[1], Rostislav E. Yavorskiy[2], and Evgeniya Vlasova[2]([✉])

[1] Skolkovo Foundation, Moscow, Russia
fk@sk.ru
[2] Higher School of Economics, National Research University, Moscow, Russia
ryavorsky@hse.ru, vlasova.eug@gmail.com

Abstract. In this paper we study formal indicators of connectivity for a city community of researchers in Informatics and Cybernetics. The analysis is based on data available at the Scientific Electronic Library portal http://eLibrary.ru. Starting from the co-authorship relation we construct connectivity graph for research institutions for all major Russian cities and suggest using size of the 2-core component to measure the connectivity of the local communities.

Keywords: Social network analysis · Research communities · Communities connectivity · Co-authorship relations

1 Introduction

1.1 Role of Collaboration in Creative Cities and Islands of Innovation

As it is stated in [1] "the *creative city* became the new hot topic among urban policymakers, planners, and economists, especially in North America and Western Europe". Needless to say that initiative of Russian government to build Skolkovo Innovation Center also follows that trend. It is stressed out in [1] that among other factors creative cities have always been associated with free exchange of scientific ideas, which naturally raises the task of developing measurable indicators of that parameter. Monograph on intelligent cities [2] mentions science and technology parks as one of five category of island of innovation and lists *collaboration* between universities and businesses as the first factor of the productive environment. So it becomes quite important to develop methods and tools for measuring the collaboration level as according to Lord Kelvin "If you cannot measure it, you cannot improve it."

1.2 How to Measure a City

In [3] fifteen indicators divided in five categories were selected to design index system of innovative city. Although the index system list contains several measures

© Springer International Publishing Switzerland 2014
D.I. Ignatov et al. (Eds.): AIST 2014, CCIS 436, pp. 111–120, 2014.
DOI: 10.1007/978-3-319-12580-0_11

of the local R&D community such as "Personal quantity engaged in R&D per million labor forces", "The proportion of R&D fund to GDP", "R&D personal full-time equivalent" and others, none of the indicators evaluates the collaboration level in the considered city.

The task of local communities ranking is quite similar to the well known problem of university ranking, see [10] and [11] for a detailed study on the subject. Another project which is closely related to our research is Map of Russian Science, which is currently in trial operation phase, see [13] and [14] for more.

Our general approach to the study of urban professional communities was described in [4]. It was based on mathematical model of a community as a dynamic socio-semantic network described in [6], see also [9]. A more detailed overview of works on scientific collaboration could be found in [7] and [8]. Some other factors of professional online communities are studied in [5].

1.3 Maturity Measures of Professional Community

In [4] an approach was proposed for assessing an urban professional IT community maturity level. It is based on measuring two sets of parameters that characterize the level of competence and the density of the network of contacts. With such approach, the formal model is described in terms of combination of social and semantic networks. The study also provides results of the pilot testing of the proposed approach for assessment of several city-wide IT-communities in central Russia.

The proposed rating of a professional community provides a system approach to assessing the current status of professional communities. However, this approach (in its current form) has a certain limited scope of applicability. First of all, a more definitive list of parameters and their weight needs to be elaborated, specifically to account for financial performance and social/demographic data of a region, information on registered legal entities, etc. In addition, a system of scoring by experts should be replaced with the one automatical, based on social/demographic and other data to minimize the human subjectivity factor.

According to [4] the key factors defining professional IT communities maturity level are competences and contacts. The both factors can be decomposed into four components. To these two groups of factors we add the third component, activity bonus, including positive factors that don't fit in the previous ones. Total 100 points:

1. Competencies (max – 40 points)
    (a) Development of IT education – 10 points
    (b) Development of IT industry – 10 points
    (c) Development of Business Education – 10 points
    (d) Research in Computer Science – 10 points
2. Contacts (max – 40 points)
    (a) Regular IT conferences and workshops – 10 points
    (b) Web communities, blogs and forums targeted at IT audience – 10 points
    (c) Groups in Social networks – 10 points
    (d) Focused IT Media – 10 points

3. Activity bonus – (max 20 points)

In this paper we assume that the academic community maturity level index can be either decomposed into the level of competence and the density of the network of contacts.

1.4   Goal of This Paper

As it was mentioned before, in the original study the scoring was performed by the group of independent experts. In this paper we suggest an automated procedure based on publicly available data on publications to measure the connectivity level of Russian major cities research computer science communities. We take into account both absolute number of scientific publications and links with the other scientific centers, which are determined as the number of co-authored papers.

2   The Dataset

Our analysis is based on data available at http://elibrary.ru, which is the largest Russian scientific portal, aggregating works on science, technology, medicine, and education. It contains over 18 million of articles and publications from more than 3200 Russian journals, see [12].

2.1   The Search Criteria

We restricted the focus of our study to Russian cities with population over 1 million citizens [15] and less than 5 to exclude cities-multimillionaires widely differing from the cities we consider in this paper by its size and structure.
    The search configuration was the following.

1. Start with the extended search form.
2. Put the city name into the search field. The Russian cities-millionaires we studied are Novosibirsk, Yekaterinburg, N. Novgorod, Kazan, Samara, Omsk, Chelyabinsk, Rostov-on-Don, Ufa, Volgograd, Krasnoyarsk, Perm, and Voronezh.
3. Indicate that the search should run through the authors affiliations, tick on all types of publications.
4. Restrict the scientific area to Cybernetics (28.00.00) or Informatics (20.00.00).
5. Switch on the morphology trigger.
6. Specify the publication years, 2011 till 2013.

Having followed these instructions we downloaded 26 collections (13 cities, 2 scientific areas) with more than half a hundred articles in each. The exact number of articles on Cybernetics and Informatics published in the last 3 years according to the Electronic Library portal is provided in Fig. 1.

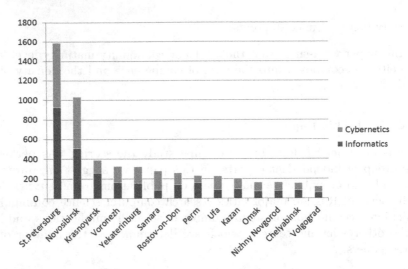

Fig. 1. Rating of major Russian cities (without Moscow) with respect to the number of 2011–2013 publications in Informatics and Cybernetics according to data from the Electronic Library portal http://eLibrary.ru

3   Research Organizations Connectivity Graph

Each paper from the data set contains information about the authors and their affiliations. At first, we extracted the list of the institutions from the considered city, whose members have published at least one research paper since 2011. As it was mentioned before, in this paper we restrict our attention to IT related areas, namely, Informatics and Cybernetics.

These institutions are the nodes of the *research organizations connectivity graph* for a city. Two institutions are connected if there is at least one paper in the data set which is co-authored by people from these institution.

Some of the papers may be co-authored by researchers from the different cities. So, to make the picture complete we also show connections to the other cities institutions on the graph. See the graph for Novosibirsk in Fig. 2.

To make the picture less noisy we remove isolated nodes and pairs. Also we stress out all the edges of the core part with bold lines. The resulting graph for Novosibirsk is given in Fig. 3.

The formal definitions follow.

We say that two different organizations v_1 and v_2 *collaborate* if there is at least one paper in the dataset co-authored by employees from the both v_1, v_2 (and maybe some other organizations). Then, *the research organizations connectivity graph* $G = (V, R)$ for a given city has the following parameters.

– V is a set of nodes. Each node denotes an organization. V consists of the following two disjoint parts:

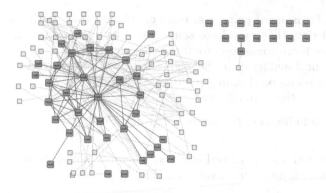

Fig. 2. Research organizations connectivity in Novosibirsk

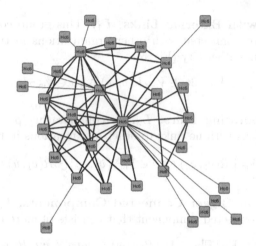

Fig. 3. Refined graph for Novosibirsk

- V_{in} comprises all organizations from the considered city such that there exists at least one paper in the dataset, which is authored or co-authored by someone working at this organization;
- V_{out} denotes all organizations outside the city, which collaborate with some organizations in V_{in}.
- R is irreflexive symmetric binary relation on V which links different collaborating organizations. So, $\forall v \neg R(v, v)$ and $\forall u, v (R(u, v) \rightarrow R(v, u))$.

4   Analysis

For each considered city all the nodes in V_{in} are naturally classified into the following six groups:

- L_0 is the subset of isolated nodes.
- L_1 denotes isolated nodes with external links.

- L_2 stands for isolated pairs, i. e. pairs of collaborating institutions from the considered city, that may have some connectors from the other cities, but cannot have more connectors from the considered one.
- L_3 stands for dangling nodes belonging to a larger connected component.
- L_4 includes nodes on dangling paths.
- L_5 nodes from the graph 2-core.

The detailed definitions of the layers are given below.

Isolated Nodes, L_0. It turned out that each city has quite a big number of organizations with no connections at all (Fig. 4).

$$L_0 = \{v \in V_{in} \mid \forall w \neg R(v, w)\} \tag{1}$$

Isolated Nodes with External Links, L_1. This group consists of institutions, which are not connected with other organizations in the city, but have collaborators in some other city (Fig. 5).

$$L_1 = \{v \in V_{in} \mid \exists w \in V_{out} R(v, w) \land \forall u \in V_{in} \neg R(v, u)\} \tag{2}$$

Nodes in Collaborating Pairs, L_2. A collaborating pair consists of two connected institutions with no links to other organizations in the city.

$$L_2 = \bigcup \{v_1, v_2 \in V_{in} \mid R(v_1, v_2) \land \forall u \in V_{in} \backslash \{v_1, v_2\} (\neg R(v_1, u) \land \neg R(v_2, u))\} \tag{3}$$

Dangling Nodes at Bigger Connected Components, L_3. Organizations linked to a larger connected component that consists of more than 2 nodes.

$$L_3 = \{v \in V_{in} \mid \exists! w \in V_{in} (R(v, w) \land \exists u (u \neq v \land R(w, u)))\} \tag{4}$$

Nodes on Dangling Paths, L_4 and the Graph 2-Core, L_5. L_4 stands for organizations linked to a connected component via one single path. This class includes nodes which have more than 1 neighbor (so they are not in L_3), but does not belong to graph 2-core, which is L_5.

Finally, L_5 is the graph 2-core, the maximal subgraph with minimum degree at least 2. This group includes all groups of several collaborating institutions with two or more collaborators each. (Sometimes 2-core is defined as a maximal connected subgraph where every node has at least two neighbours, the connectivity is not required here.) In general the connectivity analysis would require computing n-core for $n > 2$, yet for the given dataset these are empty for most of the cities.

$$L_4 = \max\{U \subseteq V_{in} \mid \forall u \in U \exists v, w \in U (v \neq w \land R(u, v) \land R(u, w))\} \tag{5}$$

And then

$$L_5 = \{v \in (V_{in} \backslash L_5) \mid \exists u, w (u \neq w \land R(v, u) \land R(v, w))\} \tag{6}$$

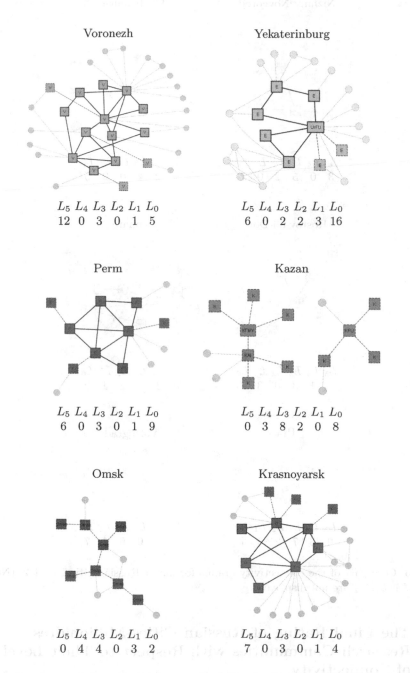

Fig. 4. Core part of the connectivity graphs for major Russian cities. Part 1. (Nodes from L_0, L_1, L_2 are not displayed.)

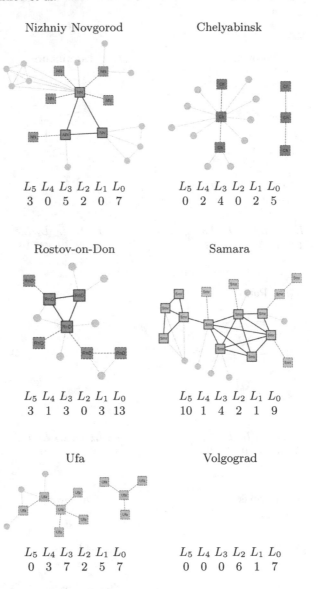

Nizhniy Novgorod

L_5 L_4 L_3 L_2 L_1 L_0
3   0   5   2   0   7

Chelyabinsk

L_5 L_4 L_3 L_2 L_1 L_0
0   2   4   0   2   5

Rostov-on-Don

L_5 L_4 L_3 L_2 L_1 L_0
3   1   3   0   3   13

Samara

L_5 L_4 L_3 L_2 L_1 L_0
10   1   4   2   1   9

Ufa

L_5 L_4 L_3 L_2 L_1 L_0
0   3   7   2   5   7

Volgograd

L_5 L_4 L_3 L_2 L_1 L_0
0   0   0   6   1   7

Fig. 5. Core part of the connectivity graphs for major Russian cities. Part 2. (Nodes from L_0, L_1, L_2 are not displayed.)

5   The Final Rating of Russian Cities-Millionaires Research Communities with Respect to Their Levels of Connectivity

The level of connectivity is one of the key points of measuring the community maturity level.

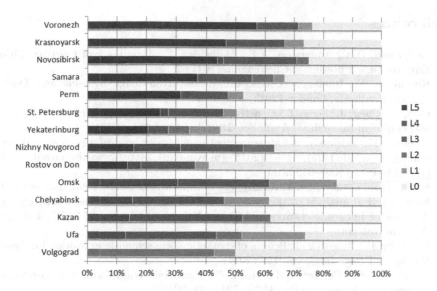

Fig. 6. The final rating of cities with respect to the connectivity of the research community based on the co-authorship relation for papers in Cybernetics and Informatics during 2011–2013

In this paper we suggest a formal procedure that helps computing the connectivity graph G for any research community, split its vertexes into several connectivity levels $L_0, \ldots, L_5$ defined above. So, each city gets the following vector of the normalized connectivity characteristics:

$$l = (l_5, l_4, l_3, l_2, l_1, l_0), \text{ where } l_i = \frac{|L_i|}{|V_{\text{in}}|} \text{ and } \sum_{i=0}^{5} l_i = 1. \tag{7}$$

According to the connectivity graphs we build the final rating of Russian cities based on lexicographical order of the vectors of normalized connectivity characteristics. See Fig. 6.

6   Conclusion

The idea of measuring urban communities was discussed in many papers and has shown its ambiguity, see e.g. [16]. Understanding the structure of urban communities considered in this paper brings us closer to finding the key phases in the development of urban communities and to the understanding of phase transitions.

Although our study at this stage aims at identifying explicit phases of development of urban communities, we believe that the observed characteristics of urban communities allow us to make a step towards building a descriptive system and individual classification of the urban communities.

References

1. Andersson, D.A., Andersson, A.E., Mellander, C.: Handbook of Creative Cities. Edward Elgar Publishing, Cheltenham (2011)
2. Komninos, N.: Intelligent Cities: Innovation, Knowledge Systems and Digital Spaces. Routledge, London (2013)
3. Guo, P., Chen, J., Song, Z.: Construction of evaluation system of innovative city and empirical research. J. Appl. Sci. **13**(22), 5428–5433 (2013)
4. Krasnov, F., Yavorskiy, R.: Measurement of maturity level of a professional community. Bus. Inf. **1**(23), 64–67 (2013) (in Russian). http://www.bijournal.hse.ru/en/2013--1(23)/86058013.html
5. Krasnov, F., Ustalov, D., Yavorskiy, R.: Comparison of online communities on the base of lexical analysis of the news feed. In: Proceedings of 2-nd Conference on Analysis of Images, Networks and Texts, Yekaterinburg, pp. 254–257, 4–6 April 2013 (in Russian)
6. Yavorsky, R.: Research challenges of dynamic socio-semantic networks. In: CEUR Workshop Proceedings, vol. 757 (2011)
7. Taramasco, C., Cointet, J.-P., Roth, C.: Academic team formation as evolving hypergraphs. Scientometrics **85**(3), 721–740 (2010)
8. Roth, C., Wu, J., Lozano, S.: Assessing impact and quality from local dynamics of citation networks. J. Inf. **6**(1), 111–120 (2012)
9. Roth, C.: Socio-semantic frameworks. Adv. Complex Syst. **16**(04–05), 1350013 (2013)
10. van Vught, F.A., Ziegele, F.: U-Multirank: design and testing the feasibility of a multidimensional global university ranking. Final report (2011). http://doc.utwente.nl/85192/1/Vught11umulti.pdf
11. van Vught, F.A., Ziegele, F.: Multidimensional Ranking: The Design and Development of U-Multirank, vol. 37. Springer, Heidelberg (2012)
12. Scientific Electronic Library. http://eLibrary.ru
13. Polyakov, A.M.: Map of Russian science. First results. Conference SCIENCE INDEX 2013: Analytical Instruments and Services for Assessment of Scientific Work. MGIMO, Moscow, 9–10 December 2013 (in Russian). http://www.elibrary.ru/projects/science_index/conf/2013/presentations/polyakov.pdf
14. Map of Russian Science. http://www.mapofscience.ru
15. Russian Federation Federal State Statistics Service. http://www.gks.ru/bgd/regl/b10_109/Main.htm
16. Cox, A.: What are communities of practice? A comparative review of four seminal works. J. Inf. Sci. **31**(6), 527–540 (2005)

A Gradient Method for Generating Facial Barcodes

Georgy Kukharev[1], Yuri Matveev[2,3]([✉]), and Nadezhda Shchegoleva[4]

[1] West Pomeranian University of Technology, Szczecin, Poland
gkukharev@wi.zut.edu.pl
[2] University ITMO, St. Petersburg, Russia
matveev@mail.ifmo.ru
[3] Speech Technology Center Ltd., St. Petersburg, Russia
matveev@speechpro.com
[4] Saint Petersburg Electrotechnical University "LETI", St. Petersburg, Russia
NLSchegoleva@etu.ru

Abstract. We propose a method for generating standard type linear barcodes from facial images. Our method uses the difference in gradients of image brightness. It involves averaging the gradients into a limited number of intervals, quantization of the results into decimal digits from 0 to 9, and table conversion into the final barcode. The proposed solution is computationally low-cost and does not require the use of any specialized image processing software for generating facial barcodes in mobile systems. Results of tests conducted on the Face94 database show that the proposed method offers a new solution for use in real-world practice. The generated barcodes are stable against changes of scale, pose and mirroring of facial images, as well as changes of facial expressions and shadows on faces from local lighting.

Keywords: Facial images · Brightness gradients · Barcodes · Mobile systems

1 Introduction

The idea of using standard barcodes for personal identification was first suggested in a 1999 patent [1]. It was assumed that personal identification would be performed at the moment a customer made an electronic payment in real time, and that the unique barcode, printed on the customer's hand or body, would be read by means of a special device. However, further real-word implementations of personal identification using barcodes were not developed in spite of the wide use of biometric identification methods. Nevertheless, barcodes are often worn today as fashionable tattoos [2]. Barcodes on the human body do not carry any biometric characteristics relating to the individual, but could be used for personal identification if they can be generated in real time, directly from a person's face or voice.

© Springer International Publishing Switzerland 2014
D.I. Ignatov et al. (Eds.): AIST 2014, CCIS 436, pp. 121–127, 2014.
DOI: 10.1007/978-3-319-12580-0_12

Assuming that such an identification procedure exists in principle, the solution may be to encode people's faces or voices [3,4] in the form of barcodes, without any permanent mark being affixed to the body. Such barcodes can be used in biometric, access control (AC) and video surveillance systems, content-based video retrieval systems, etc. However, there are some challenges to generating facial barcodes. One of them is variability (variations in lighting, pose and expression, etc.) in real-world facial images. Solving the problem would simplify personal identification and improve the performance and reliability of related recognition systems.

In this paper, we propose an approach for presenting facial images in the form of linear EAN-8, EAN-13 or UPS barcodes [5].

2   Brief Overview of Existing Approaches

Ten years after the patent [1] was published, the authors of [6] noted that facial identity was largely conveyed by horizontal image structure, such as eyebrows, eyes and lips lines. They demonstrated that this information could be successively represented as a set of binary strips or as a so-called biological barcode. Furthermore, they explored some invariant features of a person's facial biological barcode. However, as noted in [6] and in further publications by these authors, an algorithm for generating such barcodes is not defined. Instead, the authors note that facial images translated into thick, straight black and white bars will never provide an exact representation of a person's face [7].

Five years later, the most serious practical research on the problem of facial representation in the form of barcodes was published [8]. The authors proposed an algorithm for barcode generation based on searching for specific (key) points on the face, descriptions of local features, and creation of a two-dimensional color barcode. However, the algorithm is unlikely to be integrated in mobile smartphone and tablet systems in the near future due to its use of SIFT (Scale Invariant Feature Transform) and SURF (Speeded Up Robust Features) procedures for generating 2D barcodes.

The concept of a biological barcode uses an algorithm that compares two facial images, as presented in [9]. The idea of the algorithm is explained in Fig. 1. First, it calculates the brightness gradients between two specularly located bands that slide synchronously down a facial image from top to bottom, as shown in Fig. 1(a). Second, the differences between the current and mean values of the gradients are calculated and encoded. The values of the differences equal to and below zero are encoded as "0", and those above zero are encoded as "1", generating a binary code that represents a facial image similar to a biological barcode [6]. Figure 1(b) shows the current values of the gradients and their mean value; (c) shows the biological barcodes for each source image. The disadvantage of this approach is its inability to generate the same binary code for facial images of the same person when the facial images differ slightly. These differences could be insignificant but visible to the eye, such as variations in lighting, scale, facial pose and expression, etc. In order to solve this problem, we propose developing the ideas in [9, p. 240] for representing human faces in the form of standard linear barcodes.

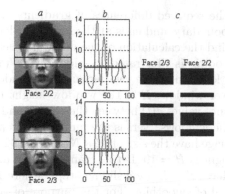

Fig. 1. Biological barcode using binary code generation

3 An Algorithm for Generating Facial Barcodes

In general the facial barcode generation system consists of four main components:
1 – image preprocessor; 2 – feature extractor; 3 – feature coder; 4 – barcode
generator.

Block 1 solves two problems. First, it analyzes the source image parameters,
including the size, color and deviation from the horizontal line of the eyes. Second, it determines the rotation of the image plane depending on the results of
the first analysis, adjusts the image size and corrects for brightness. Solutions
for these two tasks can be found in [9]. Feature extraction is implemented in
block 2 using the difference of image brightness gradients. Block 3 performs feature coding, which is the most important task. Feature coding encodes facial
features using the necessary number of decimal digits. The differences between
brightness gradients are first averaged into a limited number of intervals, and
then the results are quantized into decimal digits from 0 to 9. The task of block
4 is simply table conversion of the results of block 3. Barcode generation also
includes checksum computation for the decimal code created in block 3, and
the conversion of this code into a binary matrix that graphically represents the
source facial image as a barcode. Our approach makes it possible to generate
linear barcodes from facial images in EAN-8 format. It can also used to generate
linear barcodes in EAN-13 and UPS formats [5].

4 Generating of Facial Barcodes Based on Differences of Gradients

Feature extraction from the original image is based on calculating the difference
between brightness gradients in two mirror-located windows. The windows are
$H \geq 1$ pixels high and have the width of the original image. They slide synchronously from top to bottom across the facial image with a step of $S \geq 1$. At each
step t we calculate the distance $d(t)$ between the sub-images in the windows.

These distances are the required differences of gradients. The sliding starts at the "hair/forehead" boundary, and ends at the lower border of the nose.

The principle behind the calculation of the difference between gradients and the sliding window coding is illustrated in Fig. 2. Here (a) is the image the with initial and final location of the two rectangular windows; U is the upper window, D is the lower window, H is the window height (b) shows the curve representing the distance values $d(t)$ between the corresponding fragments of the images "covered by the windows" versus the number of steps $t = 1, 2, ..., T$.

Let the original image have the size of $M \times N = 112 \times 92$ pixels. Assume that the initial window height is $H = 10$. Let us transform the image into an EAN-8 barcode. In total we have $T = L \cdot mod$ sliding steps, where L is the code length and mod is the interval of smoothing. For the purpose of generating a barcode in EAN-8 format, the parameter $L = 7$. The parameter $mod \geqslant 8$, in general, is chosen from the condition:

$$T = L \cdot mod \leq M - H. \tag{1}$$

The value of T should fall approximately on the lower border of the nose area (or, in some cases, between the nose and lips). This will exclude the lower part of the face and thus eliminate the influence of emotion on the stability of barcode generation. Otherwise it is necessary to increase the size of the source image in block 1 until condition (1) is met.

Now we define the distance $d(t)$ between the windows:

$$d(t) = \|U(t) - D(t)\|, \forall t = 1, 2, .., T. \tag{2}$$

Results of (2) are shown in Fig. 3(a) and they are normalized in block 3:

$$d(t) = d(t)/max(d), \forall t = 1, 2, .., T. \tag{3}$$

These values are averaged over the time interval of mod, then quantized into decimal digits from 0 to 9 by means of the scale factor $scale$:

$$\bar{d}(l) = f\{scale[\sum_{j=1}^{mod} d(mod(l-1) + j)]/mod\}, \forall l = 1, 2, .., T, \tag{4}$$

where $f(.)$ is the nearest integer value; $scale$ is the scale factor $9 < scale < 10$.

The result (4) is shown in Fig. 3(b). This result is transferred from block 3 to block 4 where the final 8-digit barcode is generated. The 8th digit is the checksum for the first 7 digits from block 4. The lower part of Fig. 3(b) shows an example of the final barcode.

5   Experiments

The proposed method for generating facial barcodes was tested on images from the Face 94 database [10]. The first 100 classes with 11 images in each class were used to generate EAN-8 barcodes. We used 112×92 pixel images in GRAY format.

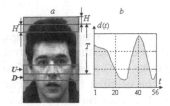

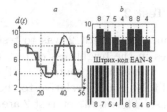

Fig. 2. Calculation of the difference between gradients and sliding window coding

Fig. 3. Normalizing and coding processes

5.1 Test 1

We generated barcodes for the images without any preprocessing and with the following coding parameters: $H = 23$; $S = 1$; $T = 56$; $L = 7$; $mod = 8$; $scale = 9.5$.

In Test 1, about 70 % of the barcodes generated for faces of the same class matched. An example from Test 1 is shown in Fig. 4. The middle column shows the barcodes derived from facial images of the same class and the phase correlation between the corresponding distance vectors (we an almost 100 % similarity). We experimented with barcode generation for facial images with different facial expressions, changes in the eyes (open or closed), mirror reflections of the original image, variations in scale, pose, as well as with shadows on facial images from local lighting. Our results clearly demonstrate that the barcode remains stable in all these cases.

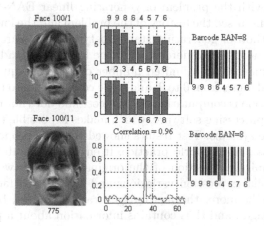

Fig. 4. Results of generating facial barcodes based on the gradient method

5.2   Test 2

The purpose of Test 2 was to test the stability of barcode generation against changes in the brightness of test images. Test 2 used the same parameters as Test 1. However, in contrast to Test 1, the brightness of the test images varied from 140 % to 60 % relative to the brightness of the original images in Face94 database. The results of Test 2 are shown in Table 1. It may be noted that when brightness changes by $\pm 20\,\%$ relative to the initial brightness, the result is stable. However, when brightness changes by $\pm 40\,\%$ relative to the initial brightness, the number of matching barcodes generated decreases by almost 50 %.

Table 1. Results of the Test 2

Test	1	2	3	4	-	5	6	7	8
Image									
Intensity	140	130	120	110	100	90	80	70	60
Number of pairs	487	584	639	744	775	746	699	560	385

6   Conclusion

This paper deals with the problem of generating linear EAN-8 barcodes from facial images. We discussed the history of the problem and the known approaches. The proposed method of generating standard type linear barcodes from facial images is based on using the difference of image brightness gradients. It involves averaging the gradients into a limited number of intervals, quantization of the results into decimal digits from 0 to 9, and table conversion into the final barcode. The proposed solution is computationally low-cost and does not require the use of specialized image processing software, which makes it possible to generate facial barcodes in mobile systems. Test results showed that the proposed method is a new solution for use in real-world practice. It ensures the stability of generated barcodes for the mirror reflection of the original image, as well as in cases of scale, pose and facial expression variations and shadows on facial images from local lighting. Furthermore, the method generates standard barcodes directly from the facial images, and thus contains information about a person's face.

Acknowledgments. This work was partially financially supported by the Government of the Russian Federation, Grant 074-U01.

References

1. Heeter, T.W.: Method for verifying human identity during electronic sale transactions. US Patent 5878155A (1999)
2. Barcode/Tattoos. http://www.barcodeart.com/store/wearable/tattoos
3. Matveev, Y.N.: Technologies of biometric identification of a person by voice and other modalities. Vestnik MGTU. Priborostroenie, Spec. Issue Biometric Technol. **3**, 46–61 (2012) (in Russian)
4. Forczmanski, P., Kukharev, G., Shchegoleva, N.: An algorithm of face recognition under difficult lighting conditions. Electr. Rev. **10b**, 201–204 (2012)
5. UPC & EAN Barcode FAQ Tutorial: http://www.idautomation.com/barcode-faq/upc-ean/
6. Dakin, S.C., Watt, R.J.: Biological bar codes in human faces. J. Vis. **9**(4), 1–10 (2009)
7. Facial Barcodes Help Us Identify People. http://www.barcodesinc.com/news/?p=92
8. Querinini, M., Italiano, G.F.: Facial recognition with 2D color barcodes. Int. J. Comput. Sci. Appl. **10**(1), 78–97 (2013)
9. Hitrov, M.V. (ed.) Methods of facial images processing and recognition in biometrics, Politechnika, Saint Petersburg (2013) (in Russian)
10. Face94 data base. http://cswww.essex.ac.uk/mv/allfaces/faces94.html

Fingerprint Identification Algorithm Based on Delaunay Triangulation and Cylinder Codes

Alexander Dremin, Mikhail Yu. Khachay, and Anton Leshko(✉)

Krasovsky Institute of Mathematics and Mechanics,
16 S. Kovalevskoy St., Ekaterinburg, Russia
anton@leshko.org

Abstract. A new efficient fingerprint identification algorithm combining a modification of the Delaunay triangulation minutiae-based hashing technique for a model dataset, the Maltonian cylinder coding fingerprint matching method, and MAP-classifier learning procedure is proposed. Numerical experiments prove the robustness of the algorithm w.r.t. small perturbations of minutiae data and the sufficiently high level of natural noising for query fingerprints. Also, performance analysis results with comparison to state-of-the-art 'Suprema' identification algorithm are presented.

Keywords: Fingerprint identification · Delaunay triangulation · Cylinder codes · Pattern recognitions

1 Introduction

For the last decades, the research and development activity in the area of automatic biometric verification and identification systems is steady increasing. A variety of biometric technologies have been proposed. Among them are fingerprints, face, iris, and speech recognition algorithms. Each technology has its own strength and shortcomings. The main criteria used for the comparative analysis of several biometric technologies are universality, uniqueness (authenticity), collectability, permanence etc.

The fingerprint biometric technology (also known as dactyloscopy) appears to be the oldest and the most popular due to its several attractable properties, among them are high personality and stability of fingerprint images. For a given finger and a given person, fingerprint is just a digital gray-scale image obtained from an optical scanner and containing a picture of papillary lines ('ridges' and 'valleys'). Thus, fingerprint verification and identification are special machine learning problems involving the development of specialized image processing, segmentation, and analysis algorithms.

It seems that the verification problem has been investigated in details, whereas the identification problem remains a great challenge for researchers and developers. Along with performance, scalability becomes one of the first-priority issues in the development of fingerprint methods.

© Springer International Publishing Switzerland 2014
D.I. Ignatov et al. (Eds.): AIST 2014, CCIS 436, pp. 128–139, 2014.
DOI: 10.1007/978-3-319-12580-0_13

Although, there are known several fingerprint verification systems, which examine raw fingerprint images using the correlation analysis techniques only, usually [1] the verification/identification stage is preceded by some feature extraction one.

Among other feature extraction methods, the minutiae-based technique is most popular. From the geometrical viewpoint, *a minutia* is an irregularity point on a fingerprint image (where termination, bifurcation, or crossover of papillary lines are observed). The collection of such points on the image plane is called *a fingerprint template*. Many different types of modern fingerprint analysis algorithms [1–3] are based on such templates.

Unfortunately, a regular fingerprint image typically contains several dozens of minutiae and the analysis of all their combinations appears to be computationally expensive. Several geometric techniques are developed [2] to reduce this combinatorial complexity, and the the triangulation-based indexing algorithm for the minutiae set is known as the most promising.

In the paper, a new fingerprint identification algorithm based on the Delaunay minutiae triangulation, special type of coding, and MAP-learning classifier is presented. The main contribution is an original feature-space construction technique based on partial invariants against some known image transformation group. Performance of the proposed algorithm is compared with performance of well-known proprietary 'Suprema' algorithm [4], which is supposed to be state-of-the-art [5].

2   Problem Statement and Related Works

2.1   Verification and Identification Problems

The are two main problems associated with biometric data: verification and identification. Verification is an one-to-one (matching) problem. The goal is to answer the question *Whether this person is who he (or she) claims to be?* Every verification system implies two stages: *enrollment* and *query*. When a system enrolls a person for the first time, in addition to the fingerprint images, some auxiliary data (e.g. name, photo, passport or driver's license id, etc.) are captured as well. When a person returns, he (or she) should present these complementary data along with the new fingerprint; the system just validates them. The verification problem is well-known. Therefore, the main goal of any research in this field, is to improve the performance of the algorithms in the following directions: fingerprint image enhancement and machine learning procedures based on detected minutiae.

Image enhancement algorithms are designed to improve the overall quality of fingerprints, thereby simplifying further minutiae detection procedures. The modern approach to fingerprint image enhancement is based on the general scheme proposed in famous paper [6] and is followed by many researchers [7–10]. According to this scheme, before the analysis, a fingerprint is segmented into regions of three types: well-defined, where ridges and valleys are clearly separable and minutiae can be easily detected; recoverable corrupted, where ridge-valley texture is corrupted but can be interpolated with a sufficiently high accuracy on the basis of neighboring areas; and unrecoverable corrupted regions. The goal is to improve the quality of recoverable regions and remove all unrecoverables. The enhancement procedure consists of the following stages: preprocessing (normalization, sharpening etc.), orientation field estimation, frequency image constructing, region mask building, and adaptive filtering (using several Gabor-like local filters).

Most modern matching algorithms [1] are centered on the geometrical alignment between previously detected minutiae (from query and model fingerprints) and constructing geometric (partial) invariants with respect to a given plane transformation group.

On the other hand, biometric identification systems answer the question *Who is this person?* The required answer should depend solely on the fingerprint image presented. At first glance, this problem can be reduced to the appropriate sequence of the verification problems. Indeed, at the enrollment stage, a hypothetic identification system can just memorize fingerprint data obtained from known people constructing so-called model database, and, at the query stage, it can search this database for most similar entries to the fingerprint in question, using some matching algorithm as a subroutine. But, this simple scheme has several shortcomings, and its poor scalability seems to be the most important.

2.2   A Structure of Identification System

From conceptual point of view, any automated fingerprint identification system (AFIS) consists of two main subsystems. First of them (we call it *offline*) is used at the enrollment stage, when a model database is constructed. The second subsystem identifies of query fingerprints on the basis of this database.

2.3   Geometric Indexing

To give a short description of the first subsystem, it is convenient to use the well-known *black box* model. By virtue of any standard minutiae extraction algorithm[1], an offline subsystem maps the initial model fingerprint set

$$\mathfrak{I} = \{I_j : \; j = 1, \ldots, N\}$$

into a family of finite subsets of $\mathbb{Z}_+^3$ (cube of the set of nonnegative integer numbers). Actually, any model image $I_j \in \mathfrak{I}$ is mapped to the subset

$$T(I_j) = \{(x_i, y_i, w_i) \in \mathbb{Z}_+^3 : \; i = 1, \ldots, N_j\}$$

that is called *a template* (see Fig. 1). For any triple, x_i and y_i coordinates define a geometric location of the i-th minutia detected on the image plane, and w_i is equal to the confidence level of this detection. So, in the beginning of the first stage, we have the set $\mathfrak{B} = \{T_j = T(I_j)\}$ of templates of the initial images (which

---

[1] We use the open-source algorithm provided by NBIS [14].

Fig. 1. Fingerprint template based on minutiae extraction

is called *a model database*). Further, to each pair (T_j, q) we assign a minutiae subset $T_{j,q} = \{(x_i, y_i, w_i) \in T_j : w_i \geq q\} \subset T_j$ consisting of minutiae filtered by their accuracy level. In the sequel, we consider projections of these subsets onto planes $H_q = \{(x, y, w) : w = q\}$ which are parallel to the coordinate plane xOy.

On the second (online) stage, the query image I is processed (in general) by the similar way, and the template $T = T(I)$ is produced, after that the final identification decision is made by the one-to-one matching T with corresponding candidates subset $\mathfrak{B}_T \subset \mathfrak{B}$ of the model database. Time complexity of this procedure (for a fixed template T) is $O(M|\mathfrak{B}_T|)$, where M is the complexity of the inner matching algorithm. So, the problem is to construct the reducing algorithm R, which to any T assigns a subset $R(T) = \mathfrak{B}_T$ satisfying the following additional constraints.

1. $|R(T)| \ll N = |\mathfrak{B}|$.
2. Let some confidence level $\alpha \in (0, 1)$ be given, and let a fingerprint I producing the query template T belong to some known person and the model database contains templates produced by another his (or her) fingerprints. Denote the subset of these templates by $\mathfrak{B}'_T$. Conditional probability P_T of the event $\mathfrak{B}'_T \cap R(T) = \varnothing$ should satisfy the inequality $P_T \leq \alpha$.

Mathematically, this problem is equivalent to the construction problem of the efficiently computable mostly powerful test statistic of the significance level α for the null hypothesis 'known person'. For any query template T, the test produces a subset $R_\alpha(T)$ of candidates for the subsequent one-to-one matching (w.r.t. T).

There are known several approaches to solve this problem. The approach based on the preliminary clusterization of the model database by the *core* type of the initial fingerprint images [1] seems to be the earliest. According to this approach, at the online stage, the query template is previously classified on the basis of its core, after that the search can be narrowed to the corresponding cluster. Unfortunately, the number of known core types is small and the distribution of the real fingerprints (among them) is far from the uniform one.

Another approach is based on indexing the model database and is supposed [2] to be more promising. Indexing procedures improve the classical two-stage identification scheme at the both stages. At the offline stage, the model database is indexed using some special hash function. At the online stage, the required subset $R(T)$ is constructed from the models with the hash values that are most similar to ones calculated from the query template T.

During the indexing substage, for any model template, several *partial invariants* (which are values of the geometrical nature that are almost invariant to a given transformation group on the plane) are computed and quantized. For instance, if some numerical features $f_1, f_2, \ldots, f_k$ of geometrical shapes of some kind formed by the fingerprint minutiae are used as partial invariants, then for any model T_i and for any shape S of interest, the record $g_1(S), \ldots, g_k(S), r_i$ is included into the indexing table. Here g_j is the quantized value of the feature f_j and r_i is a reference to the model T_i. Thus, any model template T_i is transformed to some finite subset in the k-dimensional indexing space.

The second, query stage starts with computing the same partial invariants of the template to be identified. The computed k-dimensional vectors are filtered using some system of additional constraints, which are control parameters of the algorithm. Further, the remaining vectors are used for searching in the index table and estimating the posterior probabilities for the models T_i extracted. The resulting ordered subset $R(T)$ is constructed from the most probable models according to their posterior probabilities.

Performance of indexing algorithms is suggested [2] to estimate by *correct index power (CIP)*.

Suppose, for any respondent (from a given sample), we have a pair (T_i, T_i') of fingerprints obtained from the same finger. Construct the model database $\mathfrak{B}$ from the first elements of each pair, and the test database $\mathfrak{C}$ from the second ($|\mathfrak{B}| = |\mathfrak{C}| = N$, by construction). The model $T_i \in \mathfrak{B}$ is said to be *correctly indexed* by the algorithm R if $T_i \in R(T_i')$. Let $N_{ci}(R)$ be the number of correctly indexed models, then

$$CIP(R) = \frac{N_{ci}(R)}{N}. \tag{1}$$

It is clear that $CIP(R)$ is a stochastic variable which depends, along with the algorithm R in question, on the random choice of the initial sample and the pair $(\mathfrak{B}, \mathfrak{C})$. Nevertheless, its population value can be estimated statistically on some representative fingerprint sample. In this paper, the well-known '*NIST Special Fingerprint Database 4*' [15] is used for such an estimation.

3   Our Results

We start with the description of our partial invariant data structure.

3.1   Partial Invariants

The system of invariants constructed in this paper generalizes the system proposed in [11] and extended in [12,13]. Our system contains quantities that are invariant to the rotation-translation-scaling subgroup of *similarity transforma-tion group* (on the plane). For a fixed accuracy level q of detected minutiae, to any template T, the projection $\Pi_q(T)$ of the set

$$T_q = \{(x_i, y_i, w_i) \in T : w_i \geq q\}$$

onto the plane $H_q = \{(w, y, w) : w = q\}$ is assigned and the Delaunay tri-angulation [16] of the set $\Pi_q(T)$ is constructed. The choice of the Delaunay triangulation method is due to the following reasons

(a) such a triangulation is unique for any nondegenerate finite set on the plane;
(b) the resulting triangulation consists of $O(m)$ facets, which number is sub-stantially smaller than the number $O(m^3)$ of all possible triangles with the vertices of the given m-point set;
(c) this triangulation can be constructed efficiently, we use the algorithm [17] with time-complexity $O(m \log m)$;
(d) the topological structure of the resulting triangulation is stable [18] w.r.t. small perturbations of the initial data.

Suppose, a triangle Δ is a triangulation facet with edges $a \leq b \leq c$. To this triangle, assign the vector $\nu(\Delta) = [\alpha, \beta, \gamma]$ by the formulas $\alpha = b/c$, $\beta = a/b$, and $\gamma = \cos C$ (here C is the angle opposite to the side c). This vector is invariant to any translation, rotation and scaling transform on the plane and satisfies the following inequalities

$$\frac{1}{2} < \alpha \leq 1,\ 0 < \beta \leq 1,\ -1 < \gamma \leq \frac{1}{2}.$$

Suitable discretized (particularly, to distinguish automatically isomer triangles) these parameters are used at both stages, offline and online.

3.2   Proposed Algorithm

Indexing Stage

Input.

1. Model database $\mathfrak{B} = \{T_j : j = 1, \ldots, N\}$.
2. Minimum accuracy level q for detected minutiae.
3. Maximum index values n_1, n_2, n_3.

Output. Set-valued map $h : \mathbb{Z}_q^3 \to 2^{\mathfrak{B} \times \mathfrak{D}}$ (index table) defined on integer lattice

$$[0, \ldots, n_1] \times [0, \ldots, n_2] \times [0, \ldots, n_3]$$

as follows: any triple (i, j, k) is assigned to the set of pairs (T_o, Δ_t), where T_p is some model template and the triangle Δ_t is a facet of the Delaunay triangulation of $\Pi_q(T_p)$ such that the discretized value of the vector $\nu(\Delta)$ is equal to (i, j, k).

Identification (Query) Stage

Input.

1. A query template T and minimum accuracy level q for detected minutiae.
2. Index (hash) table h.
3. Length L of the previously constructed hypothesis sample.
4. Threshold values η_1, and η_2.
5. Discretization parameters $\sigma, n_1, n_2 \ldots$ for similarity transformations.

Output. A triple (T_p, P_p, S_p), where T_p is extracted model (identification hypothesis), $P_p \geq \eta_2$ is posterior probability of its matching with T, and S_p is affine similarity transform assigning T to T_p. If there is no model satisfying $P_p \geq \eta_2$, then the query template T is *rejected*.

Scheme. The algorithm consists of two substages: pre-sampling of L most likely (to the query template) models and the final recognition.

1. Pre-sampling substage
   (a) Similar to the considerations above, to any triangulation facet Δ_t of the projection $\Pi_q(T)$ (for the query template T), an appropriate index cell (i_t, j_t, k_t) and the set $h(i_t, j_t, k_t)$ are assigned.
   (b) For any triangle $\delta_v \in h(i_t, j_t, k_t)$, an appropriate similarity transformation S_{tv} (mapping the vertices of Δ_t into corresponding vertices of δ_v) is computed. The scaling parameter λ_{tv}, the cosine $\cos \varphi_{tv}$ of the rotation angle, and the translation vector b_{tv} are discretized and the corresponding model template T_v is added as an entry to the secondary index table along with its score. To compute this score we use the angles α_i and β_i of papillary lines (w.r.t. *cores* of the initial fingerprints) at vertices of the both triangles Δ_t and δ_v previously corrected by the angle of their mutual rotation. The resulting score

$$V_v = \prod_{i=1}^{3} e^{-(\varphi(\alpha_i, \beta_i))^2 / \sigma^2},$$

   where $\varphi(\alpha, \beta) = \min\{\alpha - \beta \mod 2\pi, \beta - \alpha \mod 2\pi\}$.
   (c) Top L (according to gathered cumulative scores) hypotheses are extracted and ordered by decreasing of their scores (Fig. 2). If first two scores satisfy the condition $V_1/V_2 > \eta_1$, then the query template is *accepted* and is assigned to the first hypothesis. Otherwise the algorithm passes to the second substage.
2. Recognition substage
   (a) Let $T_1, \ldots, T_L$ be hypotheses extracted at the previous step. For each pair (T, T_i), we apply the matching algorithm [19] and compute its matching score M_i. Thus, we obtain the finite sequence

$$\mathcal{L} = ((V_i, M_i): i = 1, \ldots, L).$$

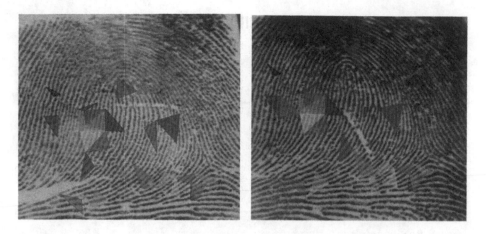

Fig. 2. Query fingerprint S1678 and the most valuable hypothesis F1678

(b) The sequence $\mathcal{L}$ is considered as an input to the previously learned MAP classifier, which searches for the hypothesis T^* being a maximizer of the estimated posterior matching probability η^*. Further, if $\eta^* \geq \eta_1$, then the query template T is *accepted* and is assigned to the hypothesis T^*; otherwise it is *rejected*.

3.3 Learning and Testing

Training (tuning control parameters) and testing of the algorithm were made on the well-known NIST-4 Special Fingerprint Database, the respectable testing source for modern fingerprint verification/identification heuristics. By structure, this dataset consists of $2\,\mathrm{K}$ fingerprint pairs, for each of them both images (denoted by '$fD_1D_2D_3D_4$' and '$sD_1D_2D_3D_4$' for some positive integer $D_1D_2D_3D_4$) are obtained twice from the same finger.

We use this dataset for solving the following additional problems.

1. Proving the stability of the proposed indexing scheme w.r.t. small perturbations of the initial data, such as addition (deletion) of minutiae and modifications of their geometrical locations.
2. Discretization parameters tuning for primal and secondary indexing tables.

According to statistical reasons, the accuracy level for detected minutiae is fixed to $q = 64$. In both problems, the subset the initial dataset consisting of 1923 (96 %) (f-image, s-image) pairs, where f-image produces a template with at least 50 minutiae, is chosen.

Proving the Stability. This kind of testing proceeds on the special synthetic dataset obtained from the mentioned above NIST-4 database. According to the well-known "white noise" model, to any f-image from the initial dataset, several

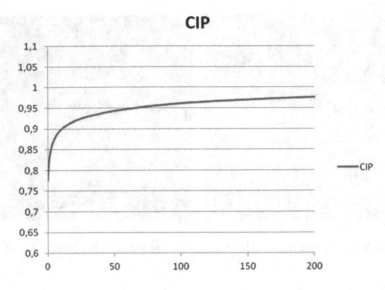

Fig. 3. CIP-index for $\sigma = 3$ and $r = 1.5$

perturbed models are assigned. For each perturbed model, the geometrical locations of the minutiae are modified by the rule $x'_i = x_i + \xi_i$, $y'_i = y_i + \eta_i$, where ξ_i
and η_i are i.i.d. $N(0, \sigma^2)$ random variables. For an additional parameter $r \in [1, 2]$,
a minutiae (x'_i, y'_i, w_i) is included to perturbed template iif $\xi_i^2 + \eta_i^2 \leq r^2\sigma^2$. Thus,
to any initial model template a 20-element perturbed sample is assigned.

Further, at the indexing stage, the initial templates are used as models
and each perturbed template is identified by the algorithm proposed. Obtained
numerical data confirm the known theoretically proved [18] stability result. Particularly, for $\sigma = 1$ and $r \in [1.5, 2]$ (from 33 % to 13 % of excluded minutiae
in average), 100 % perturbed templates are classified correctly within $L = 1$.
Increasing σ leads to increasing of the L-value, as expected. But the stability
of the entire algorithm remains high. For instance, for $\sigma = 3$ and $r = 1.5$, the
CIP-value for $L = 1$ is 77 %, and for $L = 10$ (0.5 % of the initial database), the
more then 89 % (Fig. 3).

3.4   Tuning and Final Testing

For training (parameter tuning) a subset of 430 (21%) fingerprint pairs is used,
where f-image possesses at least 100 minutiae of accuracy level 64, while the
complement of this subset (to the entire dataset) is taken as a test sample. At
the training stage, the parameters are tuned by several local search heuristics.
The optimal values of parameters are $8 \times 8 \times 8$ for the primal index table (hash)
and $17 \times 17 \times 17 \times 47$ for the secondary (Fig. 4). To learn the MAP-classifier,
the well-known k-fold cross-validation heuristic was applied.

To estimate the overall performance of the proposed algorithm, we conduct
a comparative numerical experiment on the real fingerprint dataset provided by

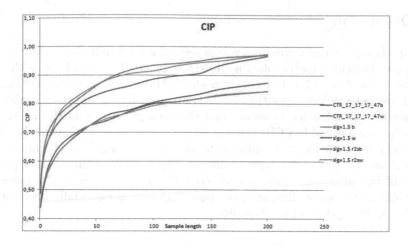

Fig. 4. CIP-analysis of several local search heuristics

Prosoft Ltd. [20]. The performance results measured for 'Suprema' algorithm (on the same dataset) are taken as a baseline (Table 1). Following the well-known approach [1] to the comparative performance evaluation of identification algorithms, for each quality level q of query fingerprint images, parameters of both algorithms are tuned so that they have the same level of *False Accept Rate (FAR)*. Further, the value of *False Reject Rate (FRR)* is used as an assessment of the algorithm's performance (for a given q).

Table 1. Comparative analysis of the proposed (the first line) and Suprema (the second) algorithms w.r.t. the quality level q of model and query fingerprints

$q > 0\%$		$q > 60\%$		$q > 70\%$		$q > 80\%$		$q > 90\%$	
FAR	FRR (%)	FAR	FRR (%)	FAR	FRR (%)	FAR	FRR (%)	FAR	FRR (%)
0,00	35,41	0,00	33,10	0,00	27,02	0,00	18,52	0,00	11,48
	35,80		30,70		26,08		22,40		18,72
0,06	33,30	0,19	30,85	0,14	26,03	0,10	18,11	0,26	10,97
	32,75		28,47		25,33		20,321		16,97
0,17	32,35	0,25	30,17	0,21	25,82	0,21	17,90	0,77	10,46
	30,20		27,69		24,53		18,47		15,54
0,35	31,76	0,43	29,80	0,28	25,32	0,31	17,49	1,02	10,20
	28,32		26,12		23,17		16,99		14,92
0,41	31,18	0,56	29,42	0,50	24,54	0,41	17,28		
	26,13		25,37		22,76		16,12		
0,94	30,12	0,99	27,31	0,64	23,97	1,03	16,36		
	25,02		24,23		20,98		15,39		

4   Discussion

A comprehensive comparative analysis of several known indexing techniques for fingerprint identification problem is presented in [2]. The algorithm with 85 % CIP value for $L = 0.1N$ is recognized the optimal among them. The indexing scheme used in this algorithm is based on considering all possible triangles with vertexes in minutiae locations, and time-complexity of its online stage is $O(Nm^3)$. Our method has CIP value of $82 \pm 5\%$ for the same L, while its time-complexity is $O(Nm \log m)$ thanks to the Delaunay triangulation technique.

As shown in Table 1 for poor quality fingerprints, the identification performance of the algorithm 'Suprema' exceeds the performance of the proposed algorithm. At the same time, the proposed algorithm is essentially better when identifying fingerprints of good quality.

5   Conclusions

A new fingerprint identification algorithm combining the Delaunay triangulation indexing, cylinder minutiae coding, and MAP-learning is presented. By the numerical evaluation it is proved that performance of the proposed algorithm is similar to 'Suprema' algorithm, which is recognized as state-of-the-art in the fingerprint identification. 'Suprema' exceeds the proposed algorithm on fingerprints of low-level quality. Therefore, it seems possible that integration of the proposed algorithm with the more advanced image enhancement techniques can improve its overall performance.

Acknowledgement. This research was supported by Russian Scientific Foundation, grant no. 14-11-00109.

References

1. Maltoni, D., Maio, D., Jain, A.K., Prabhakar, S.: Handbook of Fingerprint Recognition. Springer, New York (2005)
2. Bhanu, B., Tan, X.: Computational Algorithms for Fingerprint Recognition. Kluwer Academic Publishers, Boston (2004)
3. Mansukhani, P., Tulyakov, S., Govindaraju, V.: A framework for efficient fingerprint identification using a minutiae tree. IEEE Syst. J. **4**(2), 126–137 (2010)
4. Suprema fingerprint identification algorithm. http://www.supremainc.com/eng/technology/technology_03.php?mark=3
5. MINEX test. http://www.supremainc.com/eng/technology/technology_06.php?mark=6
6. Hong, L., Wan, Y., Jain, A.: Fingerprint image enhancement: Algorithm and performance evaluation. IEEE Trans. Pattern Anal. Mach. Intell. **20**(8), 777–789 (1998)
7. Tico, M., Vehvilainen, M., Saarinen, J.: A method of fingerprint image enhancement based on second directional derivatives. In: IEEE International Conference on Acoustics, Speech, and Signal Processing (ICASSP), pp. 985–988 (2005)

8. Nilsson, M., Dahl, M., Claesson, I.: Gray-scale image enhancement using the SMQT. In: IEEE International Conference on Image Processing, ICIP 2005, vol. 1, pp. 933–936 (2005)
9. Bartunek, J., Nilsson, M., Nordberg, J., Claesson, I.: Adaptive fingerprint binarization by frequency domain analysis. In: Fortieth Asilomar Conference on Signals, Systems and Computers, ACSSC'06, pp. 598–602 (2006)
10. Gottschlich, C.: Curved-region-based ridge frequency estimation and curved Gabor filters for fingerprint image enhancement. IEEE Trans. Image Proc. 21(4), 2220–2227 (2012)
11. Bebis, G., Deaconu, T., Georgiopulos, M.: Fingerprint Identification Using Delaunay Triangulation. In: IEEE International Conference on Intelligence, Information, and Systems, ICIIS (1999)
12. Liu, N., Yin, Y., Zhang, H.: A fingerprint matching algorithm based on delaunay triangulation net. In: Proceedings of the 2005 The Fifth International Conference on Computer and Information Technology (CIT05)
13. Liang, X., Bishnu, A., Asano, T.: A robust fingerprint indexing scheme using minutia neighborhood structure and low-order delaunay triangles. IEEE Trans. Inf. Forensics Secur. 2(4), 721–733 (2007)
14. NIST Biometric Image Software. http://www.nist.gov/itl/iad/ig/nbis.cfm
15. NIST Special Database 4. http://www.nist.gov/srd/nistsd4.cfm
16. Du, Q., Wang, D.: Recent progress in robust and quality delaunay mesh generation. J. Comput. Appl. Math. 195(1–2), 8–23 (2006)
17. Skvortsov, A.V.: A survey of algorithms fro constructing a Delaunay triangulation. Numer. Methods Program. 3, 14–39 (2002). Sect. 1
18. Guibas, L., Knuth, D., Sharir, M.: Randomized incremental construction of Delaunay and Voronoi diagrams. Algorithmica 7, 381–413 (1992)
19. Cappelli, R., Ferrara, M., Maltoni, D.: Minutia cylinder-code: a new representation and matching technique for fingerprint recognition. IEEE Trans. Pattern Anal. Mach. Intell. 32(12), 2128–2141 (2010)
20. Prosoft Ltd. http://www.prosoft.ru/company/en/

Automatic Generation of Text-Based Open Cloze Exercises

Alexey Malafeev[(✉)]

National Research University Higher School of Economics,
Nizhny Novgorod, Russia
aumalafeev@hse.ru

Abstract. This article presents an approach to the automatic generation of open cloze exercises that are based on real-life English texts. The exercise format is similar to the open cloze test used in Cambridge certificate exams (FCE, CAE, CPE). Two experiments were conducted to evaluate the usefulness on the machine-generated exercises and compare them with authentic Cambridge tests. The experiments showed that the generation method used was quite effective. With some customization, the presented method can be applied to generating similar exercises based on texts written in other languages.

Keywords: Exercise generation · Open cloze · Language exercises · Computer-assisted language learning (CALL) · English as a foreign language (EFL)

1 Introduction

Vocabulary and grammar exercises are widely used in teaching English as a foreign language (TEFL), but creating them manually is time-consuming and expensive. In response to this, many methods for the automated generation of language exercises have been proposed in the past two decades. These solutions rely on various NLP tools and techniques and can produce different types of exercises.

This paper presents an effective approach to the generation of text-based open cloze exercises similar to those used in Cambridge certificate exams (FCE, CAE and CPE). The method used is simple and does not rely on any sophisticated NLP tools, yet it is powerful enough to generate realistic and useful exercises. In fact, as shown in the evaluation section of this paper, experienced EFL instructors find it somewhat difficult to tell the difference between exercises generated with the help of the method presented here and authentic Cambridge tests.

In the most general sense, the cloze is a test of language ability or reading comprehension, which is created by removing certain words from a text. The gaps are to be filled in with appropriate words. In the open cloze, the test-taker is to guess the suitable words from the context, without seeing any multiple choice options. It may therefore be a challenging task, requiring a deep understanding of language structure [1, 2].

© Springer International Publishing Switzerland 2014
D.I. Ignatov et al. (Eds.): AIST 2014, CCIS 436, pp. 140–151, 2014.
DOI: 10.1007/978-3-319-12580-0_14

Although the open cloze may be of different varieties [2], the approach presented in this paper is aimed at the generation of exercises[1] emulating the open cloze test used in Cambridge certificate exams (FCE, CAE and CPE). In this test, "[t]he focus of the gapped words is either grammatical, such as articles, auxiliaries, prepositions, pronouns, verb tenses and forms; or lexicogrammatical, such as phrasal verbs, linkers and words within fixed phrases" [1].

The reasons for choosing the Cambridge open cloze as the target exercise type are the following:

1. Cambridge certificate exams are well-established and highly regarded, and they tend to emphasize a close relationship between teaching and testing [3]. Open cloze exercises are a useful stimulus in integrated reading, writing and vocabulary instruction [2].
2. This exercise type largely focuses on using function words in English, an analytic language. These may be difficult for learners to master and often require extensive practice. This is especially true for learners whose mother tongue (L1) differs from English in its structure. For example, Russian learners typically struggle with English prepositions, auxiliary verbs and articles, which either work very differently or are not altogether found in their L1. The open cloze may therefore be a helpful tool for practicing the use of function words.

The method discussed in this paper is part of a larger system developed by the author, called ELEM (English Language Exercise Maker), which is aimed at generating English vocabulary and grammar exercises of various types based on real-life texts. Being able to use arbitrary texts (e.g. news articles, blog entries, film reviews etc.) for generating exercises gives the user a lot of freedom in choosing interesting and relevant material. Research has shown that learners' motivation can be improved by tailoring texts to their interests [4].

This paper is structured as follows. Section 2 discusses related work. Section 3 describes the method of generating open cloze exercises used in ELEM. Section 4 reports on an evaluation of the exercises generated. Section 5 concludes and outlines future work.

2   Related Work

Some recent research has been conducted with a view to facilitating exercise creation. Among the more general solutions are multi-domain exercise or test generation systems, e.g. [5–7], as well as authoring tools, e.g. Hot Potatoes,[2] MaxAuthor[3] and others.

There are also several systems that are designed for generating exercises of one or more specific types to aid learners of the supported language(s). Exercise types include

---

[1] Or tests, depending on whether they are used for practice or assessment; for simplicity, they will be referred to as exercises.

[2] http://hotpot.uvic.ca/

[3] http://cali.arizona.edu/docs/wmaxa/

reading comprehension questions [8], morphological transformation [9–13] error correction [10], finding related words [14], shuffle questions (putting words in the right order to form a correct sentence, adding appropriate inflections and/or function words) [15], translation [16], grammatical or lexico-grammatical open cloze (fill-in-the-blank without multiple choice) [12, 13, 15], and, probably the most common, multiple choice questions or cloze tests [10, 13, 17–20]. All these are language exercises, unlike, for example, open cloze tests checking students' factual knowledge [21].

As mentioned before, this article will focus on a specific exercise type, the open cloze that tests the learner's language proficiency. The majority of the solutions listed above concentrate on other exercise types, with the exception of KillerFiller, an exercise-building tool that is part of the VISL project[4] [12], the gap-filling activity maker in the VIEW project[5] [13], and GramEx [15]. It is important to outline the principal differences between these systems and the method presented in this paper.

KillerFiller supports multiple languages including English. It extracts sentences from annotated corpora, replacing words of a given part of speech with blanks that the user has to fill in. At the time of writing, five word classes are available: verbs, adjectives, prepositions, adverbs and nouns. Open cloze exercises are only generated if the user chooses prepositions; in all other cases the lemma of the target word is given next to the blank, in which case the focus of the task is on morphological transformation. When prepositions are chosen as the target word class, they are simply removed from the sentence and replaced with gaps. Often, there is arguably not enough context for guessing correctly.

VIEW is an ICALL tool for enhancing authentic web pages in English and some other languages. It uses a blend of state-of-the-art NLP techniques: tokenization, lemmatization, morphological analysis, part-of-speech tagging, chunking, and parsing. VIEW can transform authentic web pages into language exercises. One of the supported exercise types is a fill-in-the-blank activity, where the user first selects the language phenomenon to practice (e.g. articles, determiners, gerunds vs. infinitives etc.). After providing a URL, the user proceeds to the enhanced version of the web page, where he/she is to fill the blanks made by VIEW. The system removes some of the words representing the target language phenomenon that are found on the page. The resulting activity is effectively an open cloze if the user has chosen one of the following: articles, determiners, phrasal verbs or prepositions.

GramEx is a framework for generating grammar exercises in French that are sentence-based. Unlike in KillerFiller, however, the sentences are not extracted from a corpus, but generated automatically in a strictly controlled way. Like in the two systems discussed above, the user first specifies the target word category or a more specific language phenomenon such as 'adjectives that precede the noun'. Open cloze exercises are generated if either prepositions or articles are selected. The resulting sentences are very simple, e.g. "She loves small armadillos." This may be seen as an advantage for

---

[4] Visual Interactive Syntax Learning, http://beta.visl.sdu.dk/.

[5] Visual Input Enhancement of the Web (http://sifnos.sfs.uni-tuebingen.de/VIEW/), an extension of WERTi, or Working with English Real Texts interactively (http://sifnos.sfs.uni-tuebingen.de/WERTi/).

elementary or pre-intermediate learners of French; on the other hand, higher-level learners might find the exercises too repetitive and not challenging enough.

The approach presented in this article is different in that:

1. It is developed specifically for English, and is quite language-specific. However, similar principles may be applied to the generation of this type of open cloze exercises in other languages.
2. The proposed method targets a specific exam format, and may thus be helpful in preparation for taking FCE, CAE or CPE. For many learners, the open cloze may be one of the more challenging tasks in these exams.
3. The input for the exercises is plain text files. Thus, the difficulty level of the exercise does not depend solely on the system settings (e.g. how many gaps to make or which words to remove), but also on the complexity/readability of the source text. Tools like Lexile [22] can be used to evaluate texts before generating exercises.
4. The proposed exercises are focused on words of various parts of speech, not only prepositions or determiners. Among other word classes used are conjunctions, pronouns, auxiliary and modal verbs, adverbs and particles. Furthermore, these word classes are not practiced in isolation from one another, but rather in one combined activity. This makes the task more challenging because of the larger number of gap-filling options available to the learner.
5. Although the proposed method is quite effective in generating open cloze exercises, it does not require sophisticated NLP tools such as annotated corpora (VISL), a tree adjoining grammar, syntactic and morpho-syntactic lexicons (GramEx), or even morphological analysis and part-of-speech tagging (VIEW). Therefore, it should be relatively easy to apply the method to other languages. Moreover, some NLP tools may simply not be readily available for some languages, although this is obviously not the case with English.

There is also a web-based cloze generator[6] that is somewhat similar to the proposed solution. It accepts arbitrary English texts as input and produces open cloze tests. However, much like in KillerFiller, VIEW and GramEx, it removes all words of the part of speech specified by the user, and the gaps may be very close to each other, e.g. with prepositions: *Michael Hussey picked it* ____ ____ ____ ____ *and swept that* ____ *the gap* (up from outside off; into). Although the method itself is similar (using a hardcoded list to replace certain words in the text with gaps) the resulting exercises are very different from Cambridge open cloze tests in that:

- There is always only one target word category per task;
- The same word forms can be removed from the text multiple times;
- Gaps can be too close to each other, which sometimes makes it hard, if not impossible, to restore the words.

To sum up, the presented method differs from most existing work in that it efficiently generates open cloze exercises emulating those used in Cambridge certificate

---

exams, even though it does not rely on advanced NLP tools. The next section will describe the proposed method in more detail.

3   Generation of Open Cloze Exercises

In the target type of exercise, there is a text of about 200–220 words with a certain number of gaps[7] placed at irregular intervals. Each gap is to be filled in with a single word. Consider the following example: "It is not unusual for objects only about a metre or _____ away to become unrecognizable". The blank in this sentence can be filled in with *so*, *less* or *two*. In the exam, it is enough for the candidate to give just one correct answer [1].

An empirical study of 29 FCE, CAE and CPE open cloze tests (408 gaps total) was carried out to determine what words can be removed. According to the answer keys, 198 unique word forms were accepted as correct fillers of the gaps. While a more complete list could be compiled from a larger sample of tests, the one obtained seemed representative enough. Most words on the list were function words. In fact, although forms of content words comprised about 40 % of the list, they were only used in 8 % of the gaps. This means that, roughly, in every nine out of ten cases, the test taker was to make a decision about which function word to use.

Given the dominance of function words in the chosen exercise type, it seemed possible to use a very straightforward approach to generation. It relies on a predefined set of specific word forms, rather than all words belonging to a particular class (such as prepositions) or words used in certain high-frequency combinations (collocation-based approaches). It is presumed that the target word forms can be 'safely' removed from almost any sentence in a given text. However, enough context should be provided so that the user would be able to fill in the gap with the missing word or other suitable words. Two research questions were raised:

1. Is it possible to automatically generate useful Cambridge-like open cloze exercises from English texts by relying on a static list of target word forms?
2. How similar would the resulting exercises be to open cloze tests used in Cambridge certificate exams?

Clearly, a robust list of target word forms becomes crucial for this approach. These are some key characteristics of the list:

- It should be large enough to ensure sufficient variety and difficulty of the generated exercises.
- The list should mostly contain function words: articles, prepositions, conjunctions, pronouns, particles and auxiliaries. A handful of modal verb forms can be confidently used too, provided these forms have no high-frequency homonyms in the English language. This is important because no part-of-speech disambiguation is used in the presented approach. For example, 'can' is not a very good candidate,

---

[7] 12 gaps in the FCE cloze and 15 in both CAE and CPE.

because there is a rather high-frequency homonymous noun in English. 'Could', on the other hand, is perfectly suitable.
- The list should not contain any nouns, adjectives or verbs (except for some forms of 'be', 'do' and 'have' and the already mentioned modal verb forms). A limited number of adverbs can be used, especially those that are known to be frequently misused by many learners of English.

After several months of extensive testing with EFL students at various levels of proficiency in English, 81 word forms (mainly content words) were removed and 29 (mainly function words) added to the list of 198 unique word forms obtained from the empirical study. The main criteria were word frequency, part of speech and 'restorability'. The resulting 146 forms are listed below:

a about above after again against ago all although am an and another any anybody anyone anything anywhere apart are around as at away back be because been before behind being below besides between but by could despite did do does doing done down during each either enough every everybody everyone everything few for from had hardly has have having how however if in into is it its itself just least less many more most much never no nobody none nor not nothing of off on one only onto or other others ought out over rather regardless same scarcely should since so some somebody someone something somewhere such than that the then there therefore these this those though through throughout till to too under until up was were what whatever when where whereas whether which while whilst who whose why will with within without would yet

The words on this list are known to cause learners significant difficulty, though in varying degrees. At the same time, these word forms are restorable if removed from most sentences. Thus, they can be used to make interesting and relevant gaps, i.e. ones that are neither too obvious nor too ambiguous.

The list is heterogeneous enough to provide for sufficient variety, and has mostly high-frequency words, which means that virtually any text can be used as input for exercise generation. For example, in the previous 29-word sentence thirteen of the words can potentially be removed, although, naturally, not at the same time:

(1) The list (2) is heterogeneous (3) enough (4) to provide (5) for sufficient variety, (6) and (7) has mostly high-frequency words, (8) which means (9) that virtually (10) any text can (11) be used (12) as input (13) for exercise generation.

Replacing any word of the thirteen with a gap, provided there is enough context, would result in an arguably useful open cloze question.

The exercise generation script was written in Python. It works as follows: the source text is normalized and tokenized. All words in the text that are on the above list are marked. Words of at least two characters and written in all capital letters are skipped to filter out abbreviations. A number of marked words are removed from the text at random. Note that each unique word form on the above list has an equal chance to be chosen and can only be used once per text. This is done to ensure that highest-frequency words such as articles do not dominate the exercise. Also, this is important for emulating Cambridge tests, where all gapped words in a single cloze tend to be different.

The removed words are replaced with blanks. The number of words to remove is specified at the beginning of generation. The minimum distance between the gaps is three words, irrespective of punctuation. Thus each gap has a minimum context of at least six words, which was thought to be sufficient to restore the word in most cases.

Somewhat counterintuitively, this simple and straightforward approach to the generation of open cloze exercises yielded good results. The evaluation methods used and results achieved are described in the next section.

4 Evaluation

Exercises generated with the proposed method were extensively used by language instructors with students at different levels. It was noticed that such exercises can accommodate any proficiency level, from beginner to advanced. This is largely due to the fact that the difficulty of the given exercise depends crucially on the complexity of the input text.

The final version of the solution was evaluated more formally. Two experiments were conducted to test the quality of the generated exercises and answer the research questions. The participants were expert teachers of English, all non-native speakers.

4.1 First Experiment

The aim of the first experiment was to evaluate the usefulness of individual gaps. A random sample of ten article extracts was taken from cnn.com and guardian.com. The size of each extract was about 300 words. These texts had not been previously used for exercise generation. One open cloze exercise with 12 gaps was generated from each of the sample texts using the method described in Sect. 3. The exercises were not post-edited or modified in any way.

Two expert EFL instructors with a background in preparing language students for taking Cambridge certificate exams were asked to assess the gaps in the exercises (the answer key was also provided). The instructors were to answer two questions about each of the 120 gaps:

1. Can the removed word be restored from the context?
2. Is the gap useful for teaching intermediate learners any relevant aspects of English grammar or collocation?

Table 1. Expert evaluation of the generated open cloze exercises

	Restorable gaps		Useful gaps	
	n	%	n	%
Expert 1	118	98 %	110	92 %
Expert 2	118	98 %	104	87 %
Agreement	118	98 %	106	88 %

Discussion. As can be seen from Table 1, the experts considered almost all of the removed words to be restorable from the context. As for the usefulness of the gaps, the results were predictably worse, but still rather good – about 90 %. The experts commented that most of the gaps considered 'not useful' seemed too obvious for the target proficiency level.

It seems that for the type of exercise under consideration, there might be an inverse relationship between the frequency of the removed word and the difficulty of the gap. However, further investigation might be necessary to confirm this. In any case, better results could probably be achieved by using multiple lists of removable word forms tailored to different target proficiency levels. Lists targeting lower level students should contain higher frequency word forms, and vice versa.

4.2 Second Experiment

The second experiment was aimed at comparing the generated exercises and actual Cambridge tests. Two open cloze tests from the CAE examinations in 2008 were randomly selected; these had not been previously used for compiling the list of target words. The original texts were restored by filling the gaps back in with the help of the answer keys. When two or more answers for the same gap were given in the key, the first alternative was always chosen. Open cloze exercises with 15 gaps (the same number as in the original tests) were automatically generated from the restored texts. Only one exercise variant was generated for each of the texts. As in the previous experiment, the exercises were not post-edited.

The two machine-generated and two original activities were given to two groups of EFL instructors (17 and 16 people). All instructors received identical forms with two pairs of exercises based on the two source texts. No answer keys were provided; however, most of the gaps could be easily filled in by comparing the exercises, as the majority of the gaps did not overlap.[8] The experts were told that in each pair of activities, one was machine-generated and one was an actual Cambridge test. The text formatting was identical, including the numbering of the gaps.[9] The experts were instructed to answer the same set of questions for both pairs. First, they were asked to identify the machine-generated exercise, if possible. If they felt they could do it, they were to give their answer and mark their confidence level on a scale of 0 (not sure) to 3 (very certain). Optionally, the experts were asked to specify the criteria they used to identify the machine-generated exercise.

There was no time limit for answering the questions. With the first group of 17 experts, the experiment was conducted in a university classroom. The longest it took for the experts in this group to give their answers was about 45 min. With the second group of 16 experts, the experiment was carried out by email. The results of the experiment are presented in Tables 2 and 3.

---

[8] In the first pair, 5 out of 15 gaps coincided (33 %); in the second, 2 out of 15 (13 %).

[9] In the CAE exam, the numbering of the open cloze gaps starts from 13.

Table 2. Telling the difference between automatically generated exercises and authentic tests (individual answers)

	Expert group 1					Expert group 2			
n	Answ.1[a]	Confid.1	Answ.2	Confid.2	n	Answ.1	Confid.1	Answ.2	Confid.2
1	R	2	R	n/a	1	R	3	R	1
2	R	1	W	1	2	R	2	R	2
3	n/a	n/a	R	1	3	W	1	R	1
4	R	1	R	1	4	R	1	R	1
5	R	1	W	1	5	R	2	R	2
6	n/a	0	R	1	6	W	1	R	1
7	R	1	R	1	7	R	2	W	1
8	R	n/a	W	n/a	8	W	1	W	2
9	W	2	R	2	9	R	1	R	1
10	R	2	R	2	10	R	2	W	1
11	R	2	R	1	11	R	1	R	1
12	n/a	n/a	n/a	n/a	12	R	1	n/a	n/a
13	W	n/a	W	1	13	R	1	R	1
14	W	2	R	2	14	R	1	n/a	n/a
15	n/a	0	n/a	0	15	R	2	W	2
16	W	1	R	1	16	R	1	R	2
17	W	1	n/a	n/a					

[a] R and W stand for identifying the machine-generated exercise correctly and incorrectly, respectively.

Table 3. Telling the difference between automatically generated exercises and authentic tests (aggregated results)

	Correct answers	Incorrect answers	No answer	Correct, confid. $> = 2$	Both correct	Both correct, confid. $> = 2$
n	41/66	16/66	9/66	15/66	13/33	3/33
%	62 %	24 %	14 %	23 %	39 %	9 %

Discussion. The results of the experiment show that the machine-generated exercises look quite similar to the authentic Cambridge tests. Granted, this does not imply that they are similarly useful, but it is interesting that experienced EFL instructors had considerable difficulty in differentiating between the two. Indeed, only one of the participants chose the highest confidence level for his answer (which was correct). As can be seen from Table 3, although over 60 % of the answers were correct, the instructors were not absolutely certain about their choices. Only 15 of the 66 answers were given with a confidence of 2 or higher. Remarkably, only 13 of the 33 participants were able to correctly identify both of the machine-generated exercises, and only three of them were 'almost sure' of both of their answers (confidence level 2); the other ten were less confident.

As for the optional question about the criteria used, it was answered by the experts in 46 out of the 57 cases (81 %) when one of the two activities was marked as machine-generated. However, no single criterion was consistently used for correct identification. Although the criteria used for giving 18 out of 41 correct answers (44 %) were related to the usefulness of the exercises, the same or very similar criteria were sometimes applied when making incorrect choices (6 out of 16, or 38 %). Both this fact and the low confidence levels seem to indicate that the machine-generated open cloze exercises do not look obviously less useful than the authentic ones. Admittedly, it might be necessary to perform further experiments with more exercise pairs to confirm this.

5 Conclusion and Future Work

In this paper, a simple method for generating open cloze exercises was presented. The exercises are text-based and are intended to emulate the open cloze tests used in Cambridge certificate exams (FCE, CAE and CPE). The machine-generated open cloze exercises are used in the ELEM system together with other types of language exercises focused on error correction, word formation, using verb forms etc.

The presented method relies on a list of carefully selected word forms that seem to be restorable from most contexts. Although the method does not use any sophisticated NLP tools, it seems sufficiently reliable and efficient at generating exercises of the target type, based on the evaluation described in the previous section. Presumably, the method can be used for other languages, although it might yield better results for analytic languages.

The solution described in this paper could benefit from added functionality such as checking the user's answers automatically – not simply by comparing them to the words removed from the source text, but rather by tapping into word co-occurrence data from corpora. This would make it possible to check if the word given as the answer could actually be appropriate for the context even though a different word was originally used in the source text.

Another way to continue this work could be experimenting with high-frequency content words that form numerous collocations (some examples are: 'get', 'come', 'time', etc.). Occasionally using some content words might possibly make machine-generated exercises more similar to authentic Cambridge open cloze tests. Also, it might be productive to use several different word lists to generate exercises of varying difficulty.

References

1. Cambridge English: Advanced Handbook for Teachers (2012)
2. Lee, S.H.: Beyond reading and proficiency assessment: The rational cloze procedure as stimulus for integrated reading, writing, and vocabulary instruction and teacher-student interaction in ESL. System 36, 642–660 (2008)
3. Chalhoub-Deville, M., Turner, C.E.: What to look for in ESL admission tests: Cambridge certificate exams, IELTS, and TOEFL. System 28, 523–539 (2000)

4. Heilman, M., Collins-Thompson, K., Callan, J., Eskenazi, M., Juffs, A., Wilson, L.: Personalization of reading passages improves vocabulary acquisition. Int. J. Artif. Intell. Educ. **20**, 73–98 (2010)
5. Sonntag, M.: Exercise generation by group models for autonomous web-based learning. In: 35th Euromicro Conference on Software Engineering and Advanced Applications, SEAA '09, pp. 57–63 (2009)
6. Almeida, J.J., Araujo, I., Brito, I., Carvalho, N., Machado, G.J., Pereira, R.M.S., Smirnov, G.: PASSAROLA: High-order exercise generation system. In: 2013 8th Iberian Conference on Information Systems and Technologies (CISTI), pp. 1–5 (2013)
7. Mitkov, R., Ha, A.L., Karamanis, N.: A computer-aided environment for generating multiple-choice test items. Nat. Lang. Eng. **12**, 177–194 (2006)
8. Gates, D.M.: Automatically generating reading comprehension look-back strategy: Questions from expository texts (2008)
9. Antonsen, L., Johnson, R., Trondsterud, T., Uibo, H.: Generating modular grammar exercises with finite-state transducers. In: Proceedings of Second Workshop NLP Computer-Assisted Language Learning, NODALIDA 2013, pp. 27–38 (2013)
10. Aldabe, I., de Lacalle, M.L., Maritxalar, M., Martinez, E., Uria, L.: ArikIturri: An automatic question generator based on corpora and NLP techniques. In: Ikeda, M., Ashley, K.D., Chan, T.-W. (eds.) ITS 2006. LNCS, vol. 4053, pp. 584–594. Springer, Heidelberg (2006)
11. Dickinson, M., Herring, J.: Developing online ICALL exercises for Russian. In: Proceedings of the Third Workshop on Innovative Use of NLP for Building Educational Applications, pp. 1–9. Association for Computational Linguistics, Stroudsburg, PA, USA (2008)
12. Bick, E.: Live use of corpus data and corpus annotation tools in CALL: Some new developments in VISL. Nord. Lang. Technol. Aarb. Nord. Sprogteknologisk Forskningsprogram, pp. 171–185 (2005)
13. Meurers, D., Ziai, R., Amaral, L., Boyd, A., Dimitrov, A., Metcalf, V., Ott, N.: Enhancing authentic web pages for language learners. In: Proceedings of the NAACL HLT 2010 Fifth Workshop on Innovative Use of NLP for Building Educational Applications, pp. 10–18. Association for Computational Linguistics, Stroudsburg, PA, USA (2010)
14. Heilman, M., Eskenazi, M.: Application of automatic thesaurus extraction for computer generation of vocabulary questions. In: Proceedings of the SLaTE Workshop on Speech and Language Technology in Education, pp. 65–68 (2007)
15. Perez-Beltrachini, L., Gardent, C., Kruszewski, G.: Generating grammar exercises. In: Proceedings of the Seventh Workshop on Building Educational Applications Using NLP, pp. 147–156. Association for Computational Linguistics, Stroudsburg, PA, USA (2012)
16. Burstein, J., Marcu, D.: Translation exercise assistant: Automated generation of translation exercises for native-arabic speakers learning english. In: Proceedings of HLT/EMNLP on Interactive Demonstrations, pp. 16–17. Association for Computational Linguistics, Stroudsburg, PA, USA (2005)
17. Sumita, E., Sugaya, F., Yamamoto, S.: Measuring non-native speakers' proficiency of english by using a test with automatically-generated fill-in-the-blank questions. In: Proceedings of the Second Workshop on Building Educational Applications Using NLP, pp. 61–68. Association for Computational Linguistics, Stroudsburg, PA, USA (2005)
18. Hoshino, A., Nakagawa, H.: WebExperimenter for multiple-choice question generation. In: Proceedings of HLT/EMNLP on Interactive Demonstrations, pp. 18–19. Association for Computational Linguistics, Stroudsburg, PA, USA (2005)
19. Goto, T., Kojiri, T., Watanabe, T., Iwata, T., Yamada, T.: Automatic generation system of multiple-choice cloze questions and its evaluation. Knowl. Manag. E-Learn. Int. J. KMEL. **2**, 210–224 (2010)

20. Knoop, S., Wilske, S.: WordGap - Automatic generation of gap-filling vocabulary exercises for mobile learning. In: Proceedings of Second Workshop NLP Computer-Assisted Language Learning at NODALIDA 2013, pp. 39–47 (2013)
21. Kurtasov, A.: A System for generating cloze test items from Russian-language text. In: Proceedings of the Student Research Workshop Associated with RANLP, pp. 107–112 (2013)
22. Stenner, A.J.: Measuring reading comprehension with the lexile framework. In: Fourth North American Conference on Adolescent/Adult Literacy. Washington D.C. (1996)

Alternative Ways for Loss-Given-Default Estimation in Retail Banking

Alexey Masyutin[✉]

National Research University Higher School of Economics, Moscow, Russia
alexey.masyutin@gmail.com

Abstract. The cornerstone of retail banking risk management is the estimation of the expected losses when granting a loan to the borrower. The expected losses are determined by three parameters. The first is the probability of default (PD) of the borrower. The methods of PD estimation were studied in detail by previous authors, and the most common method is credit scorecard development. The second parameter is exposure at default (EAD). Except for revolving loans, it is known in advance, it is the current balance (principal amount plus accrued interests) of the loan. Finally, there is a third parameter that defines the expected losses. This is the so-called loss given default (LGD) which is in effect the share of EAD, which is irretrievably lost in the event of default. This paper discusses several econometric techniques which allow one to obtain estimates of the LGD parameter.

Keywords: LGD · Survival analysis · Kaplan-Meier estimator · Cox regression · Beta-regression · Recovery rate

1 Introduction

Banks and financial institutions are taking credit risks in order to make profit. Despite the rigorous risk management strategies banks face considerable portion of defaulting borrowers. When the arrears occur it is necessary to carry out activities aimed to recover the significant part of the defaulted loan.

In order to build a system of effective bad debt collection banks have to discriminate among the debtors and detect the groups of those who tend to return the bigger part of the overdue amount and those who will likely give no repay at all. The problem of finding such groups is reduced to identification of significant factors influencing recoveries. Having determined which kind of borrowers pose the greatest risk of no recovery, and (what is more difficult) having obtained quantitative estimates of this risk, banking analysts are able to advise on the collection department resources allocation. If expected recovery rate is too low then bad debts can be sold to third parties. The problem is particularly relevant at the stage of the so-called late recovery (hard collection), when the bank is not limited to auto-dialing systems and instant messaging, but is forced to assign call-centers staff for direct communication with customers,

D.I. Ignatov et al. (Eds.): AIST 2014, CCIS 436, pp. 152–162, 2014.
DOI: 10.1007/978-3-319-12580-0_15

employees for personal contact with the client. In addition, the bank always faces a dilemma: to continue the collection process internally or to give it to out-sourcing. If the bank assesses the likelihood of repayment of arrears for some pool of customers as relatively high, then it is reasonable not to conclude agency agreements with collectors, because the commission for services of debt collection can reach more than 30 %. The research of overdue debts can be carried out in two directions: (1) assess the likelihood of the transition from delinquent state to healthy state, (2) evaluate the expected amount of income as a percentage of arrears (recovery rate).

2   Methods Discussion

Traditional and, apparently, the most common way to solve the first problem is to build a scorecard (collection scoring). Like in the case of PD estimation (application, behavioral scoring) the scorecard assesses the likelihood of loan full repayment. The only difference here is target variable definition. The techniques remain the same: those are relevant variables selection, variables transformation (e.g. WOE transformation) maximizing the specified criteria, and, finally, binary logit model calibration. Two important and significant limitations of the method should be mentioned. First, in order to build and validate the model there must be the evidence that an event has occurred or not (in this case, the return from the delinquent status). Therefore analyst must have a dataset containing a sam-ple of closed loans (full repayment took place) and a sample of written-off loans (loan did not return to a healthy state). Meanwhile the active loans with current delinquent status fall out of the sample, since the event of recovery (as well as no-recovery) are not yet determined. This case is highly undesirable, as soon as excluded observations also carry information. Such situation applies to the known problem of *right-censored* data. In other words, the logit model is not designed to work with censored data. This limitation is less painful for large banks, because the history of the portfolio has more than enough observations. But for small and medium-sized banks which find themselves in process of grow-ing its loan portfolio, when the maximum age of the loans reaches only 12–18 months, reduction in the sample size is impossible. Processing the censored data in this case is an inevitable necessity.

Second, collection-scoring answers the question, *how likely* this loan is going to return from the state of delinquency. However it does not aim to answer *when* it is going to recover. Therefore, time aspect remains out of focus. Indeed, scoring can answer the question, whether there will be a return within a specified period of time (usually 12 months), but at what point the transition takes place is unknown. So, collection-scoring does not assess the density distribution function of the moments of recoveries. In this paper, we use survival models (or time-to-event models) as a tool to analyze the repayment of arrears. This branch of statistics was developed in the second half of the XX century. The milestone works are the work of Edward Kaplan and Paul Meier (1958, [1]), Weibull Vallodi (1961, [2]), and David Cox (1972, [3]). Survival analysis is used primarily in the medical and sociological research. For example, one can verify the effect of certain therapy when patients are divided into those who receive treatment and those who receive placebo. In sociology this tool is used to investigate what factors influence the duration of staying unemployed. The duration of life is understood as time spent in the unemployed state, and "death" is defined as getting the job [4]. However, in credit risk management survival analysis techniques are also applicable. First of all, this is alternative way to estimate PD, when the lifetime is treated as duration of a loan without overdue (or without overdue more than X days), while the "death" refers to falling into default (Lyn C. Thomas et al. [5]). We can find other applications of survival analysis that are not connected with credit insolvency. In Stepanova et al. [8] they consider the fact of the loan early repayment. This phenomena is less painful for banks in comparison with default but still not desirable. Our approach is close to Jiri Witzany et al. [9]. However, we do not equate the LGD parameter with the survival function of the loan. We would rather consider survival function as a way to assess the proportion of the defaulted loans which are not fully recovered up to the point in time. LGD modeling is also used within small and medium enterprise (SME) segment. The relevant work is Sudheer Chava et al. [10] where LGD is estimated using survival analysis techniques. In contrast, we focus on retail segment loans, and the set of predictive variables is primarily socio-demographic characteristics of the borrower.

3   Concepts of Survival Analysis Models

Suppose we have a sample of n objects, each is defined as a random variable, i.e. lifespan: $T_1, T_2, ..., T_n$. The object is called right-censored, if its real lifespan is yet not known. For example, the object has been living for 2 years, and the researcher does not know how long it will live more. So, the observed lifespan is less of equal to the real lifespan. Thus, due to the right censoring, the researcher does not observe real lifespans $T_1, T_2, ..., T_n$, instead he observes the minimum of the real life and observed $(X_i, D_i), i = 1, 2, ..., n$:

$$X_i = min(T_i, C_i) \tag{1}$$

where C_i is a real lifespan

$$D_i = \begin{cases} 0 & \text{the subject is censored, i.e. } C_i < T_i \\ 1 & \text{otherwise} \end{cases} \tag{2}$$

But we are interested in characteristics of the initial series distribution. Namely, $f(t)$ is a lifetime density function, $F(t)$ is a lifetime distribution function, $S(t) = 1 - F(t)$ is a survival function, $h(t)$ is a hazard function. Survival function answers the question about the probability of lifetime greater than t. i.e.:

$$S(t) = Pr(T \geq t) \tag{3}$$

Hazard function (force of mortality, or the failure rate in engineering) is by definition:

$$h(t) = \lim_{dt \to 0} \frac{Pr(t \leq T < t + dt | T \geq t)}{dt} \qquad (4)$$

where T is duration of life, in our case, this is time spent in the delinquency state. It is assumed that the fact of censoring does not depend on the lifetime and performance of any subject. This is a realistic assumption, since censored subjects are in the banking collection are active loans. Under this assumption one can show how lifetime functions are related:

$$h(t) = \frac{f(t)}{S(t)} \qquad (5)$$

In its turn, the survival function can be expressed in terms of the hazard function:

$$S(t) = \exp^{-\int_0^t h(s)ds} \qquad (6)$$

Survival analysis models are aimed to obtain estimates of the last two functions.

4   Description of Data and Analysis

In our case, the object of the study is the delinquency higher than 30 days. The loan is considered to be in a state of delinquency until the borrower repays (a) the overdue principal amount, (b) overdue interests, (c) penalty for each day of delay in installments. Under the lifetime of the subject we mean duration of the delinquent status. Under the "death" of the subject we will understand full recovery, i.e. loan returning from delinquent to healthy status. It is still a question about the interdependency between consequent delinquencies within a single loan. For example, it is possible that if the loan falls into arrears for the third time, then it is highly unlikely to get any recovery. In this paper, this dependence is not modeled. Thus, if the loan had several cases of delinquency, their durations were considered to be independent random variables. In this paper we consider the auto-loan portfolio of one of the top-20 Russian banks. The loans were issued within 2010–2012 years. Due to the non-disclosure agreement the details are confidential. The sample consists of 1370 cases of delinquency. So, written-off, closed and still active loans are all presented within the dataset. If the delinquent loan is active then the observation is right-censored. Analysis of repayment of arrears was conducted within the software package STATA[1]. Despite the SAS high prevalence in the banking sector, the choice was made in favor of the package STATA due to availability of datamarts (no need to manipulate the data), as well as easier syntax of the STATA package[2].

---

[1] Seminar on time-to-event analysis is available at http://www.ats.ucla.edu/stat/stata/seminars/stata_survival/.

[2] The similar seminar but within SAS framework can be found at http://www.ats.ucla.edu/stat/sas/seminars/sas_survival/.

The variables that were significantly affecting the repayment of arrears, are given in the appendix [Table 2].

The first thing to do before going on to multivariate regressions is to check the relationship between the delinquency duration and variables separately. Non-parametric survival function estimation (Kaplan-Meier estimator) suits well for this task. The evaluation does not require any assumptions about the distribution function of the delinquency duration. The estimator assesses the probability that the duration of the installment delay exceeds t days:

$$\widehat{S(t)} = \prod_{t_i \leq t} \frac{n_i - d_i}{n_i} \tag{7}$$

where t_i is a moment in time at which the full recoveries were observed, n_i - number of loans, which preserve the delinquent status at time t_i less censored observations at this point, and d_i - the number of loans, returned into a healthy state at time t_i. Kaplan-Meier estimates can be constructed within different groups, by maturity, loan amount, and other characteristics of the borrower. Survival functions graphs can be found in appendix [Figs. 2, 3 and 4]:

Long flat tail of survival function shows that after being one year in the state of delinquency the probability of returning to a healthy state is almost zero. Further, note that the probability of exit from the delinquency state is 5–6% lower for men than for women. The same effect is observed for unmarried versus married borrowers. Parallelism of survival curves show the proportionality of risk that allows one to build multiple regression model in which the effects will be evaluated simultaneously. For this we use the semiparametric Cox model. Cox regression estimates the hazard function, suggesting that it depends on factors as follows:

$$h(t|X) = h_0(t) \exp^{X\beta} \tag{8}$$

where $h_0(t)$ is an arbitrary function (baseline hazard), X are factors and β is the vector of coefficients. The model is semi-parametric, since there is a function that is not a priori given, but on the other hand still there are parameters to be estimated. After evaluation of the extended model some insignificant variables were excluded, and eventually the model took the following form (Fig. 1):

The coefficients in column *Haz. Ratio* show how many times the hazard function will increase when the regressor in its turn increases by one. Since many variables (such as education, type of loan, sex) are discrete, the coefficients indicate how the hazard functions differ between the groups. So, if the loan is of the third type, at any point in time "risk" of exiting the delinquency state is up to 2.53 times higher compared to the first type of loan. If the borrower has more than one child, the "risk" to leave the state of delinquency increases in 1.14 times. The difference between the borrower with higher education and complete secondary is 1.3 times. Alternative way of lifetime estimation is presented by the parametric methods. The most popular is the use of the lognormal distribution function, as well as the Weibull distribution. Parameters distributions are estimated within likelihood maximization method. For brevity, we give a report on a model constructed for the lognormal distribution:

```
. stcox age sex_enc i.education_enc credit_sum credit_period num_depend i.type

        failure _d:  vyshel
   analysis time _t:  prosrochka_1

Iteration 0:   log likelihood = -4057.4778
Iteration 1:   log likelihood = -4024.0101
Iteration 2:   log likelihood = -4023.1607
Iteration 3:   log likelihood = -4023.1553
Iteration 4:   log likelihood = -4023.1553
Refining estimates:
Iteration 0:   log likelihood = -4023.1553

Cox regression -- Breslow method for ties

No. of subjects =      1237               Number of obs   =      1237
No. of failures =       615
Time at risk    =    223963
                                          LR chi2(11)     =     68.64
Log likelihood  =  -4023.1553             Prob > chi2     =    0.0000
```

_t	Haz. Ratio	Std. Err.	z	P>|z|	[95% Conf. Interval]	
age	.9879588	.004548	-2.63	0.008	.9790849	.9969132
sex_enc	.768267	.0681181	-2.97	0.003	.6457148	.9140788
education_~c						
2	.97061	.1416051	-0.20	0.838	.7292232	1.2919
3	1.3043	.1245625	2.78	0.005	1.081649	1.572781
4	1.106318	.1964495	0.57	0.569	.7811452	1.566853
5	2.839574	1.293924	2.29	0.022	1.16247	6.936249
credit_sum	.9999993	1.65e-07	-3.99	0.000	.999999	.9999997
credit_per~d	1.00508	.0027373	1.86	0.063	.9997293	1.010459
num_depend	1.138421	.0523611	2.82	0.005	1.040285	1.245815
type						
2	.4741113	.0854275	-4.14	0.000	.3330481	.6749221
3	.3952999	.0819458	-4.48	0.000	.2633131	.5934457

Fig. 1. Cox regression model output in STATA

The distribution of the delinquency duration is defined as follows:

$$\ln T \sim N(X\beta, \sigma) \qquad (9)$$

Coefficient sign shows the direction of delinquency duration change, due to the change in corresponding factor per unit. Coefficients themselves can be used to construct the probability of loan remaining in delinquency state in the next predetermined time interval:

$$Pr(T > t_0 + \delta t | T > t_0) = \frac{N(\frac{X\beta}{\sigma} - \frac{1}{\sigma}\ln t_0 + \delta t)}{N(\frac{X\beta}{\sigma} - \frac{1}{\sigma}\ln t_0)} \qquad (10)$$

t_0 is time (in days) since loan has fallen into delinquency. Using this formula, for instance, one can calculate the probability of remaining in a delinquent state within the next 30 days for specific borrowers.

5   Recovery Rate Estimation

To solve the second problem, set at the beginning of the work, namely recovery rate estimation, we will use the beta-regression (Silvia et al. [6]). However, there is a sufficient restriction of the beta regression: censored data cannot be processed. The density function of the random variable having a beta distribution is defined as follows:

$$f(z, \beta, \gamma) = \frac{\Gamma(\beta + \gamma)}{\Gamma(\beta) + \Gamma(\gamma)} z^{\beta - 1}(1 - z)^{\gamma - 1} \tag{11}$$

And in case of regressors:

$$f(z, \beta, \gamma, X) = \frac{\Gamma(X\beta + X\gamma)}{\Gamma(X\beta) + \Gamma(X\gamma)} z^{X\beta - 1}(1 - z)^{X\gamma - 1} \tag{12}$$

In this formula β and γ are now coefficient vectors. The distribution is used in the analysis of a continuous variable strictly limited by 0 and 1. Since the problem of estimating the probability of full recovery from delinquent to healthy state was considered in the first part, we will focus on those loans, which showed only partial recovery. Its distribution is bimodal around zero and unity: that means that in most cases the debtor either completely pays back, or nothing at all (Table 1).

Table 1. Beta regression output in STATA

betafit rr, alphavar(perc_sum credit_period) betavar(age credit_period type1)						
Iteration 0: log likelihood = 111.88367						
Iteration 1: log likelihood = 159.03024						
Iteration 2: log likelihood = 175.59134						
Iteration 3: log likelihood = 177.08649						
Iteration 4: log likelihood = 177.1083						
Iteration 5: log likelihood = 177.10831						
ML fit of beta (alpha,beta)			Number of obs = 531			
			Wald chi2(2) = 19.38			
Log likelihood = 177.10831			Prob >chi2 = 0.0001			
rr	**Coef.**	**Std. Err.**	**z**	**P>|z|**	**[95% Conf.**	**Interval]**
alpha						
perc_sum	.4156041	.131536	3.16	0.002	.1577983	.6734098
credit_period	-.0122402	.0028874	-4.24	0.000	-.0178994	-.006581
_cons	1.013151	.1503694	6.74	0.000	.7184326	1.30787
beta						
age	-.018639	.0062506	-2.98	0.003	-.0308899	-.0063881
perc_sum	.9988718	.2704875	3.69	0.000	.468726	1.529018
credit_period	.0120325	.0054593	2.20	0.028	.0013324	.0227326
type1	-.0161823	.1245794	-0.13	0.897	-.2603534	.2279889
_cons	.9595355	.3603647	2.66	0.008	.2532337	1.665837

The report shows that in our sample, there are 531 observations with recovery rate strictly within the limits of zero and one. If full and zero recovery cases are added then the sample makes a total of 900 observations. It is less than 1370 delinquency cases in the original sample because beta regression does not handle censored data. When estimating the parameters of these observations have to be ruled out. Only significant variables were left in the regression model. Coefficients are used to build up the distribution of recovery rate (rr), conditional on the characteristics of the loan:

$$Pr(rr \leq z|X) = \int_0^z f(s, \hat{\beta}, \hat{\gamma}, X)ds \qquad (13)$$

From this the expected recovery rate for the loan can be determined as:

$$E(rr|X) = \int_0^1 sf(s, \hat{\beta}, \hat{\gamma}, X)ds = \frac{X\hat{\beta}}{X(\hat{\beta} + \hat{\gamma})} \qquad (14)$$

For example, one can analyze the differences in the context of the loan types. Indeed the first type loans pose a greater risk of low recovery, rather than the third [Table 3]:

6 Conclusion

We provided the analysis of bad debts in terms of the temporal structure of recoveries. The probabilities of full repayment can be assessed within different groups of borrowers with the help of non-parametric methods such as Kaplan-Meier estimators. They not only help to avoid any assumptions about the density functions but also visualize the result in a comprehensive way. More detailed interconnections between repayments and borrowers characteristics are revealed by Cox proportional hazards model. Finally, the loans which showed only partial recovery were analyzed with beta regression. The described techniques allow one to discriminate bad debtors into groups with high versus low recovery rates. This provides the instrument for efficient debt collection process, when bank focuses primarily on the defaulted borrowers who are likely to pay back. Meanwhile, the portfolio of bad loans with low expected recovery rate can be sold to third parties and collection agencies. In our further research we would like to analyze factors that determine recoveries using concept-based learning [7].

Acknowledgments. Author would like to express his gratitude to Ivan Medvedev, Head of Retail Risks at RN Bank (former RCI Banque representative office) for being a guide in the world of banking risk management.

Appendix

Table 2. Variables influencing the recovery rate

Variable	Block	Description	Possible values	Transcript
Age	Borrower	Age	N	In full years
sex_enc		Sex	1	Male
			0	Female
marital_status_enc		Marital Status	1	Single/Married
			0	Other
education_enc		Education	1 (B)	Complete secondary education
			2 (C)	Incomplete higher
			3 (D)	Higher
			4 (E)	Two or more higher
			5 (F)	Academic degree
num_depend		Number of dependents	N	
is_estate_enc		Availability of real estate owned	1	Yes
			0	No
company_type_enc	The employer	Type of company	0	Without state participation
			1	With public participation
company_age		Age of	N	
company_count_staff		Number of employees	N	
credit_sum	Loan	Amount of credit	N	In the rub
credit_period		Term	N	In months
perc_sum		Principal debt / original loan amount	0 to 1	
Type		Specific loan classification adopted in the bank	1	Not associated with the loan price parameters
			2	
			3	

Table 3. Differences in recovery rate between two types of loan

	Client X	Client Y
age	25	25
credit_period	24	24
perc_sum	20 %	20 %
Type	1	3
Expected rr	**22.6 %**	**44.4 %**

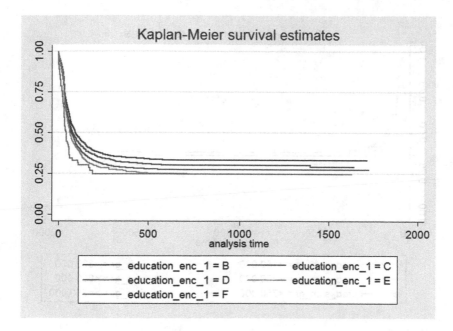

Fig. 2. Survival function within groups by education

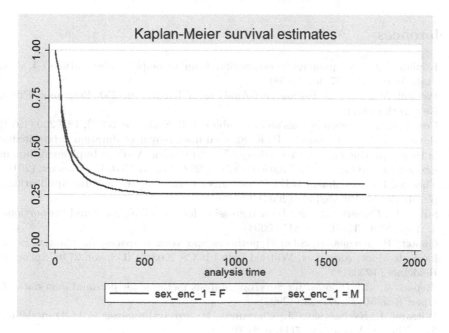

Fig. 3. Survival function within groups by sex

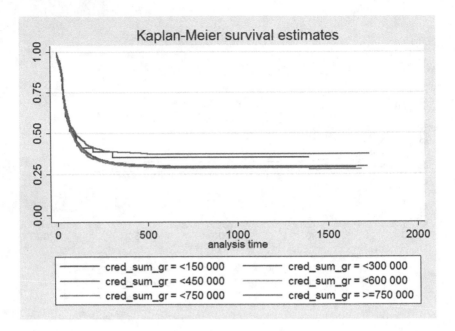

Fig. 4. Survival function within groups by loan amount

References

1. Kaplan, E.L.: Nonparametric estimation from incomplete observations. J. Am. Stat. Assoc. **53**, 457–481 (1958)
2. Weibull, W.: Fatigue Testing and Analysis of Results, p. 250. Pergamon Press, New York (1961)
3. Cox, D.R.: Regression models and life-tables. J. R. Stat. Soc. **34**(2), 187–220 (1972)
4. Ratnikova, T. A., Furmanov, K. K. Factual unemployment duration determinants in Russia (in Russian) In: Proceedings of <<Statistical Analysis Implementation in Economics and Quality Estimation>>, pp. 202–206, NRU-HSE – Moscow (2010)
5. Thomas, L.C., Edelman, D.B., Crook, J.N.: Credit Scoring and Its Applications, p. 250. SIAM, Plliladelphia (2002)
6. Silvia, F., Cribari-Neto, F.: Beta regression for modelling rates and proportions. J. Appl. Stat. **31**(7), 799–815 (2004)
7. Ganter, B., Kuznetsov, S.O.: Hypotheses and version spaces. In: Ganter, Bernhard, de Moor, Aldo, Lex, Wilfried (eds.) ICCS 2003. LNCS, vol. 2746. Springer, Heidelberg (2003)
8. Stepanova, M., Thomas, L.: Survival analysis methods for personal loan data. J. Oper. Res. **50**(2), 277–289 (2002)
9. Witzany, J., Rychnovsky, M., Charamza, P.: Survival analysis in LGD modeling. Eur. Financ. Account. J. **7**(1), 6–27 (2012)
10. Chava, S., Stefanescu, C., Turnbull, S.: Modeling the loss distribution. J. Manage. Sci. **57**(7), 1267–1287 (2011)

Conceptual Maps: Construction Over a Text Collection and Analysis

Egor N. Morenko, Ekaterina L. Chernyak$^{(\boxtimes)}$, and Boris G. Mirkin

National Research University - Higher School of Economics,
Applied Mathematics and Informatics School, Moscow, Russia
{emorenko,echernyak,bmirkin}@hse.ru
http://ami.hse.ru/vitext

Abstract. A method for conceptual maps construction is presented and applied to Business domains. A conceptual map is a graph, where nodes stand for domain specific concepts and edges connect associated concepts. The conceptual map reveals and visualises logical associations between concepts, which exist in the collection of texts, used to construct the conceptual map. Preliminary work on conceptual map analysis is suggested.

Keywords: Conceptual map · Text analysis · Annotated suffix tree · Text visualization

1 Introduction

Computational methods for text analysis are one of the main tasks of modern computer science. Logical associations between key phrases is one of the most efficient methods for solving these tasks. We construct logical associations according to their cooccurrence frequencies in the text collection. We use a method, developed in the "Method of text analysis and visualization" research group under guidance of prof. B. Mirkin (grant 13-05-0047 of Academic Fund Program). This method outputs a graph of relations between key phrases. Such a graph is a sort of conceptual graphs introduced in [1]. A conceptual graph is a directed graph, where every node is labeled with a concept presented by a key phrase. The edges of the conceptual express logical associations. In our case, the edge from node A to node B shows that the concept A most likely implies the concept B. Let us address such kind of graphs as conceptual maps.

Given the collection of texts, we can construct a conceptual map. This map will represent relations in this collection [2,3]. Conceptual maps in a number of domains:

- exploration tool of student essays and tests [4];
- text representation model used in categorisation task [5];
- text visualisation tool [6].

© Springer International Publishing Switzerland 2014
D.I. Ignatov et al. (Eds.): AIST 2014, CCIS 436, pp. 163–168, 2014.
DOI: 10.1007/978-3-319-12580-0_16

Our conceptual maps have general purpose: they show concepts and logical associations between them and can be considered as a visualisation of text collection. By now we limit our experiments only to case of one-to-one logical associations, since:

- even this oversimplified case gives enough material for examination and interpretation;
- this case is easier for visualisation.

Below the method for conceptual maps construction and several problems in work are presented.

2   Method for Conceptual Map Construction

A conceptual map is built for the twofold dataset. It consists of the list of key phrases and the text collection, taken from the same domain. We consider only those key phrases that are formulated by a human expert. Existing unsupervised methods of multiword key phrase extraction give rather poor results. Supervised methods require some efforts and are not the point of this research [7,8]. Key phrases represent concepts, which are important for a specific domain. The method is based on the procedure of scoring "key phrase to text" relevance. In this work we use the annotated suffix tree (AST) scoring to compute key phrase to text relevance in the same fashion as it is presented in [9]. The main construction is the "key phrase to text" table, organised as in Table 1. The strings are key phrases, the columns are texts and the relevance values are put in the cells. If the relevance value is lower than the given threshold, we suppose the text is not about this particular key phase. Otherwise, if the relevance is higher than the given threshold we decide that the content of the text corresponds to the key phrase. Usually we set up the relevance threshold at the level of 0.2, which makes up around a third part of the maximum experimental AST relevance value.

Table 1. The template key_phrase-to-text table

	$Text_1$	$Text_2$	...	$Text_m$
$Keyphrase_1$				
$Keyphrase_2$				
...				
$Keyphrase_n$				

Given the relevance threshold we define the set of texts, which are relevant for every key phrase. Let us denote key phrases as $k_i, i = 1 : n$, and let $F(k_i)$ be the set of texts, relevant to the key phrase k_i. Let us consider that the key phrase k_i implies the key phrase k_j, if the number of object which belong both to $F(k_j)$ and $F(k_i)$ makes out a significant part of i $F(k_j)$:

$$\frac{|F(k_j) \cap F(k_i)|}{|F(k_j)|} > r$$

where r is a confidence threshold and belongs to the $(0.5, 1]$ interval.

Thus we get the structure of logical associations between key phrases and draw the conceptual map, where nodes are key phrases and edges are logic associations. Our method reminds of the associative rules framework. However there are important distinctions between them. Associative rules are discovered in the whole dataset and require two types of thresholds, confidence and support. Since we search for connections between pre-defined key phrases, there is no need in the support threshold.

3   Conceptual Maps of Business Domain

The input dataset consists of:

1. The collection of Russian Wikipedia articles, which belong to Business category
2. The set of key phrases, describing current business situation in Russia (for more details, see [9]) (in Russian)

Wikipedia as the Source of Input Data. The Internet encyclopedia Wikipedia is one of the most popular sources of textual data in applied researches. Wikipedia is being constantly improved, updated and corrected from mistakes and is covered by free content licenses. These facts make Wikipedia so much in demand. We developed a utility that resolves some issues, that everyone faces when extracting data from Wikipedia, such as cycles in the category tree or little semantic connection of subcategories to parental categories. The utility works interactively and extracts the fragments of the Wikipedia category tree. A user enters a starting category. Than the utility lists all subcategories of this category. It checks whether there is an already traversed subcategory. If so, this subcategory is excluded from the list. The user selects necessary categories and the utility traverses them. It stops working when no subcategory is selected. The utility outputs the category tree and preprocessed articles from traversed categories.

Wikipedia-Based Conceptual Map "Business". Firstly we decided to map every key phrase to a Wikipedia category, but latter we discovered that some key phrases are too precise and there are not enough articles. So we rested content with 6686 articles, collected from the "Business" category. We set up the relevance threshold at 0.1, since the average relevance value was lower than 0.2 and lowered the confidence threshold to 0.7. Such low relevance values were achieved because of the length of the key phrases that do not occur in Wikipedia

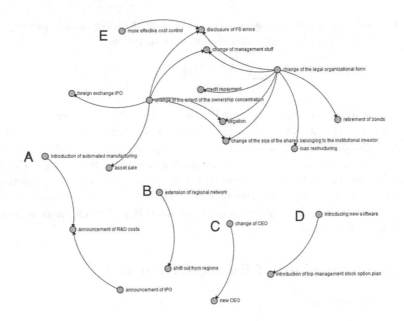

Fig. 1. The conceptual map of "Business" domain

articles. The resulting conceptual map is presented on the Fig. 1. The Figure is achieved by means of D3 library [10].

This conceptual map is a bipartite graph of five connected components. It has 21 nodes. Let us take a look on some of connected components.

Component A. Announcement of IPO requires of any kind of costs, including R&D costs. Automating manufacturing is not possible without R&D and makes R&D costs unavoidable.

Component B. Making cost control more effective requires increasing control over the whole firm or revision of business development plan; both of these possibilities are not possible without disclosure of financial statements errors.

Component C. The first level node, "change of the extent of the ownership concentration" generalises 5 concepts of second level. Perhaps to change of the extent of the ownership concentration one should take some of these five actions.

Component D. This component presents a tautologism: "change of CEO" and "new CEO".

Component E. This part of conceptual map shows possible consequence of change of the legal organisational form.

On the whole, the conceptual map might be interpreted as "action-consequence" or "goal-action" binary relation. It shows, what consequence entail some action, what actions should be taken to achieve certain goals. For example, the action "introduction of automated manufacturing" is followed by the consequence "announcement of R&D costs" or "more effective cost control" is the action, "disclosure of FS errors" is the consequence. Let us recall that the conceptual map presents the personal views of Wikipedia users, which are sometimes contradictive and discordant.

Newspaper-Based Conceptual Map "Business". Let us compare the "Business" conceptual mapped, constructed over the set of Wikipedia articles (Fig. 1), to another "Business" conceptual map (Fig. 2) of the same key phrases, constructed over the set of news messages from business newspapers.

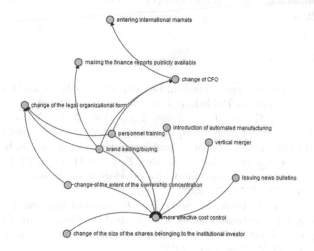

Fig. 2. The second conceptual map of "Business" domain

The structure of both conceptual maps looks similar: they do not have cycles and both are almost bipartite graphs. However, no chains coincide. The newspaper conceptual map has one connected component, only 12 nodes. The Wikipedia conceptual map is more disjoint: it has 5 connected components, comprises 21 nodes. There are more nodes of the upper level in the newspaper conceptual map, and more nodes of the lower level in the Wikipedia conceptual map. We suppose, that the maps should be interpreted in different ways: the logical associations in the newspaper-based conceptual map mean "source-target" relations, while the logical associations in the Wikipedia-bases conceptual map stand for "action-consequence" or "goal-action" relations. The conceptual map constructed over the Wikipedia articles is more of informative kind as the Wikipedia itself. The conceptual map constructed over the Wikipedia presents the real life picture. To sum up, we have two conceptual maps that comprise the same set of concepts, but are constructed over different text collections. Despite of having some visual similarities they have complete different meaning. It means that conceptual map is indeed a method for analysing the content of the text collection.

4   Conclusion

In whole, construction and analysis of conceptual maps seems to be an interesting idea, to which more attention should be paid. A conceptual map visualises not only key phrases, but also relations of key phrases inside a huge amount of texts. The subjects of further development are: automatising of setting both relevance and confidence thresholds and considering "one to many", "many to one", "many to many" types of association. The way the conceptual map can be properly interpreted is also an issue to face. To interpret the conceptual map we need to: (a) adopt several methods for analysing complex graphs; (b) develop a method for comparing conceptual maps; (c) choose a method for automated key phrase generation.

References

1. Sowa, J.F.: Knowledge Representation: Logical, Philosophical, and Computational Foundations. Brooks Cole Publishing Co., Pacific Grove (2000)
2. Tseng, S.S., Sue, P.C., Su, J.M., Weng, J.F., Tsai, W.N.: A new approach for constructing the concept map. Comput. Educ. **49**(3), 691–707 (2007)
3. Ifenthaler, D.: Relational, structural, and semantic analysis of graphical representations and concept maps. Educ. Tech. Res. Dev. **58**(1), 81–97 (2010)
4. Hwang, G.-J.: A conceptual map model for developing intelligent tutoring systems. Comput. Educ. **4**(3), 217–235 (2003)
5. Mishra, M., Huan, J., Bleik, S., Song, M.: Biomedical text categorization with concept graph representations using a controlled vocabulary. In: Proceedings of the 11th International Workshop on Data Mining in Bioinformatics (BIOKDD '12), pp. 26–32. ACM, New York (2012)
6. Ranwez, S., Ranwez, V., Villerd, J., Crampes, M.: Ontological distance measures for information visualisation on conceptual maps. In: Meersman, R., Tari, Z., Herrero, P. (eds.) OTM 2006 Workshops. LNCS, vol. 4278, pp. 1050–1061. Springer, Heidelberg (2006)
7. Liu, F., Pennell, D., Liu, F., Liu, Y.: Unsupervised approaches for automatic keyword extraction using meeting transcripts. In: Proceedings of Human Language Technologies: The 2009 Annual Conference of the North American Chapter of the Association for Computational Linguistics (NAACL '09), pp. 620–628. Association for Computational Linguistics, Stroudsburg (2009)
8. Hulth, A.: Improved automatic keyword extraction given more linguistic knowledge. In: Proceedings of the 2003 Conference on Empirical Methods in Natural Language Processing (EMNLP '03), pp. 216–223. Association for Computational Linguistics, Stroudsburg (2003)
9. Mirkin, B.G., Chernyak, E.L., Chugunova, O.N.: The method of annotated suffix trees for scoring the extent of belongingness of string to text documents. Bus. Infromatics **3**(21), 31–41 (2010)
10. Data-Driven Documents (D3.js). http://www.d3js.org. Accessed 20 January 14

Detecting Gender by Full Name: Experiments with the Russian Language

Alexander Panchenko[1,2](✉) and Andrey Teterin[1]

[1] Digital Society Laboratory LLC, Moscow, Russia
alexander.panchenko@uclouvain.be
[2] Université catholique de Louvain, Louvain-la-Neuve, Belgium

Abstract. This paper describes a method that detects gender of a person by his/her full name. While some approaches were proposed for English language, little has been done so far for Russian. We fill this gap and present a large-scale experiment on a dataset of 100,000 Russian full names from Facebook. Our method is based on three types of features (word endings, character n-grams and dictionary of names) combined within a linear supervised model. Experiments show that the proposed simple and computationally efficient approach yields excellent results achieving accuracy up to 96 %.

Keywords: Gender detection · Short text classification

1 Introduction

The Web is full of user-generated content: a plethora of platforms and technologies let a user create comments, posts and other types of textual messages. Most of the time, some information about author of a given text is available. In its simplest form, an author is represented with a *name string*, being either a real name or an alias. While some platforms, such as Facebook and Google+, let the user indicate gender, age and other information, others, such as Twitter or Web forums do not. Furthermore, in most of the platforms, only a name string is required, while other fields, such as gender, can be left unspecified. This is why often only two things can be used to describe a user: her name string and her texts.

However, in many cases it is desirable to know more about an author of a given piece of user-generated content. For instance, in the Internet marketing information about gender and age helps to improve targeting of advertisements [1]. In cyber security, user profiling can help track down Internet predators and assist in investigations of crimes [2]. In information retrieval, sociodemographic attributes can help customize user search experience and provide more relevant results [3,4].

All these factors motivate the need for systems that infer gender, age and other latent sociodemographic attributes of a user. In this paper, we investigate

© Springer International Publishing Switzerland 2014
D.I. Ignatov et al. (Eds.): AIST 2014, CCIS 436, pp. 169–182, 2014.
DOI: 10.1007/978-3-319-12580-0_17

one particular task in this research direction – *gender detection*. In particular, we focus on gender recognition of Russian full names. The goal of our method is to guess gender of a person by her name. Such a technology can be of use when a true user name is known, but gender was not specified, e.g. for analysis of Twitter users.

There have been some attempts to propose a method for automatic gender recognition (see Sect. 2). Major limitations of the prior researches are following: (i) most of the studies focus on gender recognition using text written by a person, neglecting full name of a user; (ii) most of the prior works deal with English language, neglecting particularities of other languages, such as Russian. Indeed, most researchers have focused on gender detection by text [2,5–10] with some exceptions, such as [11]. Accuracy of the state-of-the-art approaches in this field is about 80–90 %. However, in practical tasks higher accuracy is desirable. We show that one can recognize gender of a person with accuracy higher than 95 % if a full name is available (it is normally the case in social networks and blogs).

Our work fills the gaps mentioned above, as we study gender recognition methods for Russian language based on full name of a person. The main contributions of this paper are as follows. First, we show that for Russian language, the problem is not very difficult. Even a simplest statistical model yields accuracy of 85 %. Second, we propose a more sophisticated, yet simple and efficient, method that is able to recognize gender of a Russian name with accuracy and precision up to 96 %.

It is possible that some Russian Internet companies, such as Yandex[1] and Mail.ru[2] already developed technologies similar to ours [12]. However, to the best of our knowledge, we are the first to openly describe details of such technology for the Russian language. A live demo of our method is freely available online[3].

2   Related Work

Gender of a text author is often known. This makes it easy to build a training corpus of articles, blogs or posts labeled with gender tags. Provided that the gender detection technology has immediate applications ranging from marketing to cyber-security, no wonder many researchers tried to build supervised models predicting gender by text.

Koppel et al. [5] describe an approach that identifies gender of an author of a written document. In this experiment a genre-labeled subset of the BNC corpus[4] was used. The proposed method relies on combination of lexical and syntactic features and yields accuracy of roughly 80 %. The best performance in this experiment was achieved by a linear model based on function words and parts-of-speech n-grams.

---

[1] http://www.yandex.com/
[2] http://www.mail.ru/
[3] http://research.digsolab.com/gender
[4] http://www.natcorp.ox.ac.uk/

Goswami et al. [6] describe a gender detection experiment with 9,660 gender-labeled blog posts from the blogger.com platform. Features used in this experiment include slang words statistics, sentence length and other stylometric parameters. The proposed model yields accuracy of 89.3 %.

Mukherjee and Liu [13] propose two novel approaches to gender classification. The first is based on variable length POS sequences, while the second relies on automatic feature selection. The authors report increase in accuracy due to these features from 79.6 % to 88.6 % on a collection of 3,100 gender-labeled blogs from the blogger.com.

Peersman et al. [2] describe a method for short text classification by gender. The authors deal with a gender-balanced corpus of messages coming from the Dutch social network *Netlog*. The proposed technique relies on an SVM classifier [14] with features based on word/character unigrams, bigrams and trigrams. The best accuracy score of 88,8 % in this experiment was achieved with a model based on 50,000 most informative word unigrams selected with χ^2 test.

Daniel and Zelenkov [12] performed a statistical analysis of the spoken sub-corpus of the Russian National Corpus[5]. It appeared that in public communication there is a statistically significant difference between the speech of men and women (men talk more), while the same difference is absent in private communication. The article also mentions a gender recognition system for written texts with accuracy "about 90 %" trained on the same corpus.

It is worth mentioning that age and gender prediction have much in common. First, often researchers tackle two these problems in the same study [2,6,7]. Second, the state-of-the-art techniques for age and gender prediction are fairly similar. In their simplest form these methods are based on supervised linear models trained on character and/or lexical unigrams. Nguyen et al. [8] also points out an interaction between age and gender variables. Furthermore, the authors study impact of gender on quality of age prediction. For instance, it was found that age prediction works better for females.

Ciot et al. [9] were among the first to present an experiment on non-English data. The authors tackled the gender recognition problem of French, Japanese, Indonesian and Turkish texts. This study revealed that (i) the methods working well in English yield good results as well for French, Turkish and Indonesian; (ii) baseline methods provide poor results for Japanese; (iii) language-specific features can boost accuracy of the baseline approach.

The work of Burger et al. [11] is arguably the one most similar to our research. The authors proposed a model based on features extracted both from texts written by a person and his full name. The study is based on a dataset of 184 thousand Twitter users speaking more than 13 languages. However, Russian-speaking users were not studied in this experiment. Similarly to other researchers, Burger et al. used supervised models trained on character and word n-grams. Their model based on full names achieved an accuracy of 89 %, while the model based on all text fields of Twitter profile provided an accuracy of 92 %.

---

[5] http://www.ruscorpora.ru/en/

Thus, recently gender detection technology received a significant attention in the literature. Some further related experiments include Rao et al. [10], Rangel and Rosso [7], Al Zamal et al. [15] and Lui et al. [16].

3 Dataset

We performed our experiments on a dataset of 100,000 names of Facebook users with publicly available profiles. Each such *name string* contains first and last name of a user or whatever information the user inserted into this field instead. Each name string has a gender label: *male* or *female*. We did not consider users with unknown gender. The dataset was collected with the Facebook API[6] from publicly available profiles of Russian-speaking users. The dataset contains both names written in Cyrillic and Latin alphabets, e.g. "Alexander Ivanov" and its Cyrillic equivalent "Александр Иванов".

Figure 1 lists the most frequent names and surnames in the dataset. Here and below we present transliterated versions of Cyrillic characters. In our experiment, we considered the first token of a name string as a given name and the second one as a surname. It is clear from the table that in Russian language: (i) information about gender is encoded in the endings; (ii) there is a gender agreement. For instance, "Alexandr Ivanov" is a man's name, "Alexandra Ivanova" is a woman's name, and "Alexandr Ivanova" is an ungrammatical name.

While our dataset represents a significant number of common Russian names, some rare names are under-represented. For instance, the female name "Oksana Kim", is a relatively rare, but a perfectly valid female name. Yet, it is not present in our dataset (see Fig. 1). However, there are 745 people with the first name "Oksana" and 70 persons with the last name "Kim".

In order to assess difficulty of the gender recognition task, we analyzed endings of first and last names in a sample of 10,000 objects. In this context, an *ending* is a substring composed of last two characters of a first/last name. According to this experiment, 72 % of first names and 68 % of surnames have typical male/female ending. Here a *typical male/female ending* is an ending that splits males from females with an error less than 5 % (see Table 1). It appears that gender of more than 50 % given names from our sample can be robustly detected with 8 endings. Furthermore, gender of more than 50 % second names can be recognized with only 5 endings (see Table 1). These observations suggest that a simple symbolic ending-based method cannot robustly classify about 30 % of names. This motivates the need for a more sophisticated statistical approach.

4 Gender Detection Method

The gender detection method takes as input a string representing a name of a person and outputs a gender (male or female). A name string is usually extracted from a user profile. Thus, we tackle the problem as a binary classification task.

---

[6] https://developers.facebook.com/tools/explorer

		264	251	167	131	130	128	126	117	116	115	115	106	105	96	94	92	89	89	88	83	81	81	76	74	71	70
		Ivanova	Ivanov	Kuznetsova	Kuznetsov	Vasilyeva	Smirnov	Smirnova	Petrov	Shevchenko	Popova	Petrova	Popov	Bondarenko	Morozova	Volkova	Novikova	Sokolova	Mihailova	Vasilyev	Kovalenko	Romanova	Pavlova	Andreeva	Kravchenko	Alekseeva	Kim
3193	Aleksandr	0	25	0	13	0	16	0	11	7	0	0	16	6	0	0	0	0	0	12	4	0	0	0	4	0	4
2650	Elena	19	0	11	0	11	0	13	0	3	9	7	0	7	5	11	11	4	5	0	3	5	7	3	3	4	2
2620	Sergey	0	20	0	6	0	13	0	5	1	0	0	5	11	0	0	0	0	0	9	6	0	0	0	2	0	0
2222	Tatyana	12	0	10	0	10	0	9	0	7	8	11	0	0	13	4	4	9	5	0	1	0	6	4	3	5	2
2174	Olga	19	0	14	0	12	0	7	0	2	7	6	0	2	7	7	4	5	0	0	4	6	2	3	1	0	3
1976	Andrey	0	16	0	10	0	11	0	8	3	0	0	7	1	0	0	0	0	0	3	2	0	0	0	1	0	1
1914	Irina	16	0	6	0	5	0	8	0	0	5	7	0	1	3	4	4	10	2	0	2	8	3	6	2	3	1
1895	Natalya	14	0	13	0	6	0	4	0	1	5	5	0	4	9	3	6	2	7	0	1	3	3	5	2	2	1
1793	Aleksey	0	13	0	7	0	6	0	10	1	0	0	7	4	0	0	0	0	0	1	1	0	0	0	1	0	1
1721	Dmitry	0	14	0	8	0	8	0	3	5	0	0	8	4	0	0	0	0	0	4	1	0	0	0	0	0	0
1576	Svetlana	12	0	6	0	6	0	4	0	1	5	5	0	0	1	6	10	4	3	0	1	1	4	2	2	5	1
1449	Vladimir	0	13	0	5	0	4	0	7	1	0	0	2	5	0	0	0	0	0	2	0	0	0	0	3	0	4
1399	Yulia	4	0	9	0	3	0	7	0	4	1	0	0	1	0	1	2	2	3	0	3	1	1	1	0	3	2
1348	Anna	10	0	7	0	6	0	7	0	0	3	6	0	2	3	1	0	7	5	0	0	4	3	4	0	1	2
1216	Ekaterina	8	0	5	0	5	0	5	0	5	1	3	0	2	4	5	4	5	5	0	3	3	3	2	0	2	0
1199	Marina	8	0	5	0	5	0	4	0	0	6	5	0	1	4	5	2	3	4	0	1	1	4	3	2	4	3
1154	Evgeny	0	8	0	3	0	4	0	3	3	0	0	7	4	0	0	0	0	0	4	1	0	0	0	2	0	2
945	Igor	0	6	0	4	0	3	0	4	2	0	0	1	2	0	0	0	0	0	3	0	0	0	0	1	0	1
920	Anastasiya	5	0	7	0	5	0	3	0	1	0	1	0	0	2	3	3	2	1	0	1	6	0	0	3	2	0
857	Mariya	7	0	0	0	1	0	2	0	0	3	4	0	1	3	1	3	1	1	0	0	2	6	1	0	2	0
846	Oleg	0	5	0	3	0	5	0	2	2	0	0	3	0	0	0	0	0	0	1	1	0	0	0	1	0	2
822	Mihail	0	8	0	2	0	5	0	3	2	0	0	1	0	0	0	0	0	0	2	1	0	0	0	1	0	0
783	Ludmila	5	0	5	0	4	0	3	0	3	0	0	0	1	3	4	2	1	3	0	0	3	3	3	2	2	0
745	Oksana	5	0	1	0	1	0	0	0	3	2	3	0	1	2	1	1	1	2	0	4	0	3	0	0	1	0

Fig. 1. Name-surname co-occurrences: rows and columns are sorted by frequency.

A label with unknown gender can be obtained by means of the reject option [17, p.42]. Below we describe features and the model used in our gender recognition approach. This section is concluded with a description of a simple rule-based baseline.

4.1 Features

In our experiments, we used three types of features based respectively on endings, character n-grams and a dictionary of male/female names and surnames.

Word Endings. As we already mentioned above, Russian language, unlike English, has a gender agreement. Thus, the same name or surname often has different endings for a male and a female:

– males: Alexander Yaroskavski, Oleg Arbuzov
– females: Alexandra Yaroskavskaya, Nayaliya Arbuzova

Thus, some Russian surnames are transliterated differently for males and females (see above). However, other surnames are spelled in the same way for both genders, e.g. "Sidorenko", "Moroz" or "Bondar".

The two most common one-character endings of female names/surnames are "a" and "ya" ("я" in Cyrillic). We use four features that indicate on female gender in Russian language: (1) first name ends with "a", (2) first name ends with "я", (3) last name ends with "a", (4) last name ends with "я".

Table 1. Most discriminative and frequent two character endings of Russian names.

Type	Ending		Gender	Error, %	Example
first name	na	(на)	female	0.27	Ekaterina
first name	iya	(ия)	female	0.32	Anastasiya
first name	ei	(ей)	male	0.16	Sergei
first name	dr	(др)	male	0.00	Alexandr
first name	ga	(га)	male	4.94	Serega
first name	an	(ан)	male	4.99	Ivan
first name	la	(ла)	female	4.23	Luidmila
first name	ii	(ий)	male	0.34	Yurii
second name	va	(ва)	female	0.28	Morozova
second name	ov	(ов)	male	0.21	Objedkov
second name	na	(на)	female	2.22	Matyushina
second name	ev	(ев)	male	0.44	Sergeev
second name	in	(ин)	male	1.94	Teterin

Character n-grams. These features rely on character unigrams, bigrams or trigrams extracted from the name strings. We represent a name with k its most frequent n-grams. The extraction is done with help of the NLTK module [18].

Most frequent trigrams are listed below (here "_" denotes a beginning or an end of a name string): a__, va_, v__, na, ova,__A, ov_, ina, kov, nov, _Al, n__, a__,o__,__V, ndr, Ale, iya, ei, lek, eks, ko_, nko, rin,_An, enk,__C, na_, "ii ",__E, eva,__N,__M, san, ksa,__I, ev_, in_, and, len,__O, va_, rov, _Na. Note that the most frequent trigrams are not necessarily the most discriminative, e.g. "ko_" is a common surname ending for both males and females.

Dictionaries of First and Last Names. This type of features is based on dictionaries of first and last names. Each entry of a dictionary contains a first/last name and a probability that it belongs to the male gender:

$$P(c = male|w) = \frac{n^w_{male}}{\sum_{c \in \{male, female\}} n^w_c},$$

where n^w_{male} is a number of male profiles with the first/last name w in the dictionary. We use two dictionary-based features: (1) probability that first name is of male gender $P(c = male|firstname)$, (2) probability that the last name is of male gender $P(c = male|lastname)$.

We used 90,000 name strings to build the two dictionaries. The training set used in our experiments does not contain any of these 90,000 samples. In order to remove noisy entries, we deleted all names and surnames that occurred only once. The full dictionary of first names contained 3,427 entries, while the dictionary of last names contained 11,411 entries. We used several versions of these full dictionaries in our study. Each version included top γ % most frequent given names and surnames.

4.2   Model

In all our experiments we used L2-regularized *Logistic Regression* model [19] as it generally yields reasonable results for the NLP-related problems [20]. Let $y_i \in \{-1, +1\}$ be a gender label and $\mathbf{x} = (x_1, \ldots, x_n)$ be a set of features representing a name string. The logistic regression combines features in a linear combination with weights $\mathbf{w}$. This weight vector is obtained by minimizing the following unconstrained optimization problem [21]:

$$\min_{\mathbf{w}} \sum_i log(1 + e^{-y_i \mathbf{w}^T \mathbf{x}_i}) + \frac{1}{C} ||\mathbf{w}||_2$$

We use the `scikit-learn` module in this experiment[7]. This implementation relies on the Dual Coordinate Descent Method for training of the model [22]. Default meta-parameters of the model were used. The model uses nearly no regularization – the inverse of the regularization strength C was set to 100,000. Optimization of the meta-parameters, such as C, or using more sophisticated models, such as SVM [14] can lead to significant improvements in results. However, in this paper we focus on a comparison of different features used withing the framework of one model.

The model $\mathbf{w}$ can be applied to perform classification of a name string represented with a feature vector $\mathbf{x}$ as follows:

$$P(y = 1|\mathbf{x}) = \frac{1}{1 + e^{-\mathbf{w}^T \mathbf{x}}}.$$

4.3   Rule-Based Baseline

There exist several available spelling dictionaries of Russian male and female first names, such as:

- a reference dictionary of personal names[8];
- a dictionary of personal names of Russian language by F. L. Ageenko[9];
- category "Names" of Russian Wiktionary[10];
- Russian spelling dictionary of Wikisource[11].

These dictionaries can be used to classify full names by gender. We compiled a dictionary of 1,428 Russian first names labeled with gender from Wiktionary and Wikisource[12]. This dictionary has no names assigned to both male and female categories. The dictionary was used to implement a rule-based baseline that works as follows. First, an input name string is transformed into a set of

---

[7] http://scikit-learn.org/
[8] http://imena-list.ru/
[9] http://www.gramota.ru/slovari/info/ag/
[10] http://ru.wiktionary.org/wiki/Категория:Имена
[11] http://ru.wikisource.org/wiki/Орфографический_словарь_русского_языка
[12] Available at http://panchenko.me/gender/wiki-gender-dict.csv.

Table 2. Results of the experiments on the training set of 10,000 names (10-fold cross-validation). Here *endings* – 4 Russian female endings, *trigrams* – 1000 most frequent 3-grams, *dictionary* – name/surname dictionary with $\gamma = 80\%$ top entries. This table presents precision, recall and F-measure of the female class.

Model	Accuracy	Precision	Recall	F-measure
rule-based baseline	0,638	**0,995**	0,633	0,774
endings	0,850 ± 0,002	0,921 ± 0,003	0,784 ± 0,004	0,847 ± 0,002
3-grams	0,944 ± 0,003	0,948 ± 0,003	0,946 ± 0,003	0,947 ± 0,003
dicts	0,956 ± 0,002	**0,992 ± 0,001**	0,925 ± 0,003	0,957 ± 0,002
endings+3-grams	0,946 ± 0,003	0,950 ± 0,002	0,947 ± 0,004	0,949 ± 0,003
3-grams+dicts	0,956 ± 0,003	0,960 ± 0,003	0,957 ± 0,004	0,959 ± 0,003
endings+3-grams+dicts	**0,957 ± 0,003**	0,961 ± 0,003	**0,959 ± 0,004**	**0,960 ± 0,002**

tokens t. Let d_f be a set of female first names and d_m be a set of male first names. Second, gender c is assigned to a full name with the following rule:

$$c = \begin{cases} male, & \text{if } (t \cap d_m \neq \emptyset) \text{ and } (t \cap d_f = \emptyset), \\ female, & \text{if } (t \cap d_m = \emptyset) \text{ and } (t \cap d_f \neq \emptyset), \\ unknown, & \text{else}. \end{cases}$$

Thus, a person with a female first name will be considered as a female, while a person with an unknown first name will have no gender label.

5   Results and Discussion

In this section, we present results of the experiments with the gender classification approaches described above. We start with the results of the rule-based baseline. Next, we proceed to statistical models based on one type of features: endings, character trigrams and dictionary. Finally, we present results of the statistical models that rely on several kinds of features at the same time.

Table 2 presents main results of our experiments in terms of the standard performance metrics calculated on a sample of 10,000 name strings. Among these 10,000 names we found 127 (1.27 %) duplicate names. Figure 2 illustrates how size of the training set affects accuracy.

5.1   Rule-Based Baseline

As one can see, the rule-based classifier is very precise. It achieves precision of 0.995. However recall of this method is only 0.663. This is due to a large number of unclassified examples: the dictionary used by this classifier does not list many first names common on Facebook.

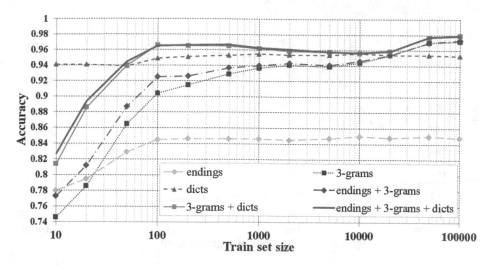

Fig. 2. Learning curves of single and combined models. Accuracy was estimated on separate sample of 10,000 names.

5.2 Word Endings

As to the statistical classifiers, even the simplest model that relies on two female endings "a" and "ya" ("я") yields reasonable results, achieving accuracy up to 0.850. However, while precision of such a naive model is relatively high (0.921), its recall is only 0.784. Therefore, a significant fraction of female names do not follow simple ending-based rules. Naturally, performance of the ending-based model improves a little as training set grows (see Fig. 2). A hundred of examples is sufficient for training.

5.3 Character n-grams

In these experiments we focus on trigrams, as according to our results they worked better than bigrams and unigrams. Table 2 and Fig. 2 present performance of the models based on 1,000 most frequent trigrams. Character trigrams significantly outperform word endings if a training set is larger than 30 samples. Furthermore, unlike the ending-based model, the trigram-based model improves as the training set grows reaching accuracy of 0.944 on a dataset of 10,000 samples.

Figure 3 plots accuracy of the *3-grams* model function of the number of most frequent trigrams used. In our further experiments (the combined models e.g. *3-grams+dicts*) we used a model based on top 1000 trigrams as a good trade-off between computational complexity and accuracy.

5.4 Dictionary of First and Last Names

The model *dicts* that relies on a dictionaries of given names and surnames yields very competitive results (accuracy up to 0.956). Furthermore, this model

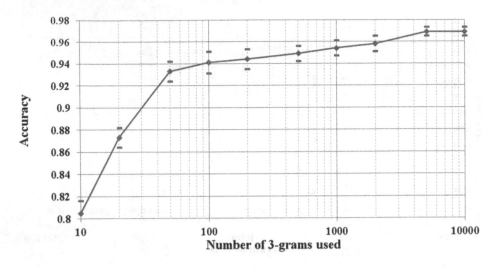

Fig. 3. Accuracy of the model *3-grams* function of the number of features used k.

provides the best precision among all statistical models (0.992). However, its recall is significantly lower than that of trigrams and combined models. As one may expect, only several dozens of training examples are enough to train this model (see Fig. 2). Further increase of the training set naturally does not improve accuracy as (i) it has only two features; (ii) the dictionaries are extracted from an independent part of data, not from the training set.

Figure 4 plots accuracy of the *dicts* model function of the dictionary size γ. The maximal accuracy is achieved if the full dictionary is used ($\gamma = 100\,\%$). However, in our experiments we used $\gamma = 80\,\%$ of the dictionaries (2,741 first names and 9,128 last names) as the difference between the two respective dictionary-based models is minimal. In fact, even if one would use only $\gamma = 60\,\%$ of the dictionaries (2,056 first names and 6,846 last names) one will get nearly the same results.

5.5   Combined Models

We experimented with three combined models: *endings+3-grams*, *3-grams+dicts* and *endings+3-grams+dicts*. Combination of trigrams with word endings yields nearly the same performance as the model based on trigrams only. Indeed, the two female endings "a" and "ya" ("я") are present among frequent trigrams, such as "a__ ", "va_", "na_", "ya__", "aya_", or "iya_".

On the other hand, the model that relies on both dictionary and trigrams outperforms both trigram- and dictionary-based models, reaching accuracy of 0.956 due to increase in recall. However, precision of this combined model is significantly lower than that of the single dictionary-based model (0.960 versus 0.992). This is so as an n-gram model can capture noisy sequences or estimate poorly weights of some n-grams due to sparsity of the training data.

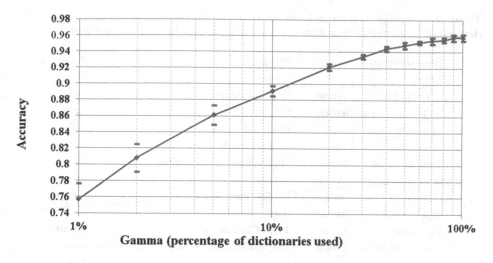

Fig. 4. Accuracy of the *dicts* model function of the fraction of dictionaries used γ.

Finally, the model *endings+3-grams+dicts* that makes use of all three types of features shows slightly higher performance than *3-grams+dicts*. However, this difference is not statistically significant. As we already discussed above, trigrams model well word endings. Table 3 lists some errors produced by our best model *endings+3-grams+dicts* on test and train sets. Train set error is very small (about 0.3 %), while test set error is bigger (about 4 %). There are several types of errors:

- Inconsistent annotation, such as "Anna Kryukova (male)" or "Boris Krolchansky (female)".
- Name string is neither male nor female, but rather a name of a group, e.g. "Wikom Tools", "Kazakh University of Humanities" or "Privat Bank".
- Name string represents a foreign name, e.g. "Abdulloh Ibn Abdulloh", "Brooke Alisson", "Ulpetay Niyetbay" or "Yola Dolson". Our model was not trained to deal with such names.
- Meaningless or partially anonymized names, e.g. "Crazzy Ma", "Un Petit Diable", "Vv Tt", "Vio La Tor" or "Muu Muu". Additional information is required to derive gender of such users.
- People with rare names or surnames, e.g. "Guldjan Reyzova", "Yagun Zumpelich" or "Akob Saakan". These are people with common names and surnames in other countries, such as Georgia, Kazakhstan, Azerbaijan or Tajikistan. In our dataset, people from these countries are under-represented.
- Full names that can denote both males and females, e.g. "Jenya Chekulenko", "Jenya Sergienko", "Sasha Sidorenko" or "Sasha Radchenko". Additional information is required to infer gender of such people.
- Misclassifications of common names, e.g. "Ilya Nasorshin", "Oleg Dubovik" or "Elena Antropova".

Table 3. Examples of train and test set errors of the model *endings+3-grams+dicts*. Here Cyrillic characters were transliterated into English in the standard way.

	Train set errors Name	True class	Test set errors Name	True class
1	Lea Shraiber	female	Ilya Nadorshin	male
2	Profanum Vulgus	female	Rustem Saledinov	male
3	Anna Kryukova	male	Erkin Bahlamet	male
4	Gin Amaya	male	Gocha Lapachi	male
5	Gertrud Gallet	female	Muttaqiyyah Abdulvahhab	female
6	Dolores Laughter	female	Yola Dolson	female
7	Di Nolik	male	Heiran Gasanova	female
8	Jlija Hotieca	female	Hadji Murad	male
9	Gic Globmedic	female	Jenya Chekulenko	female
10	Ulpetay Niyetbay	female	Tury.Ru Domodedovskaya Metro Office	male
11	Olga Shoff	male	Elmira Nabizade	female
12	Phil Golosoun	male	Niko Liparteliani	male
13	Tsitsino Shurgaya	female	Oleg Grin'	male
14	Anna Grobov	female	Santi Zarovneva	female
15	Linguini Incident	female	Misha Badali	male
16	Toma Oganesyan	female	Che Serega	male
17	Swon Swetik	female	Petr Kiyashko	male
18	Adel Simon	female	Sandugash Botabaeva	female
19	Ant Kam-	male	Jenya Sergienko	female
20	Xristi Xitrozver	female	Abdulloh Ibn Abdulloh	female
21	Anii Reznookova	female	Naikaita Laitvainenko	male
22	Aurelia Grishko	male	Fil Kalnitskiy	male
23	Alex Bu	female	Helen Hovel'	female
24	Karen Karine	female	Valery Kotelnikov	male
25	Russian Spain	female	Max Od	male
26	Lucy Walter	male	Jean Kvartshelia	male
27	Aysah Ahmed	female	Adjedo Trupachuli	female
28	Kiti Iz	female	Ainur Serikova	female
29	Cutejilian Juka	female	Privat Bank	female
30	Azer Dunja	male	No Limit	female

6   Conclusion

We presented several simple and computationally efficient models for gender detection by a full name of a person. The methods yields excellent results on a dataset Russian-speaking Facebook users, achieving accuracy up to 96 %. In our further research, we plan to complement the developed approach by a gender detection method based on texts written by a person, as a full name is not always available or can give no clue about gender. For instance, the name "Sasha Sidorenko" can refer to both a male and a female.

Acknowledgments. This research was supported by Digital Society Laboratory LLC. We thank Kirill Shileev, Segei Objedkov and three anonymous reviewers for their helpful comments that significantly improved quality of this paper.

References

1. Underwood, A.: Gender targeting for promoted products now available, October 2012
2. Peersman, C., Daelemans, W., Van Vaerenbergh, L.: Predicting age and gender in online social networks. In: Proceedings of the 3rd International Workshop on Search and Mining User-Generated Contents, pp. 37–44. ACM (2011)
3. Kharitonov, E., Serdyukov, P.: Gender-aware re-ranking. In: Proceedings of the 35th International ACM SIGIR Conference on Research and Development in Information Retrieval, pp. 1081–1082. ACM (2012)
4. Bi, B., Shokouhi, M., Kosinski, M., Graepel, T.: Inferring the demographics of search users: social data meets search queries. In: Proceedings of the 22nd International Conference on World Wide Web, International World Wide Web Conferences Steering Committee, pp. 131–140 (2013)
5. Koppel, M., Argamon, S., Shimoni, A.R.: Automatically categorizing written texts by author gender. Literary Linguist. Comput. **17**(4), 401–412 (2002)
6. Goswami, S., Sarkar, S., Rustagi, M.: Stylometric analysis of bloggers age and gender. In: Third International AAAI Conference on Weblogs and Social Media (2009)
7. Rangel, F., Rosso, P.: Use of language and author profiling: Identification of gender and age. In: Natural Language Processing and Cognitive Science, p. 177 (2013)
8. Nguyen, D., Gravel, R., Trieschnigg, D., Meder, T.: how old do you think i am: a study of language and age in twitter. In: Proceedings of the Seventh International AAAI Conference on Weblogs and Social Media (2013)
9. Ciot, M., Sonderegger, M., Ruths, D.: Gender inference of twitter users in non-english contexts. In: Proceedings of the 2013 Conference on Empirical Methods in Natural Language Processing, Seattle, Wash, pp. 18–21 (2013)
10. Rao, D., Yarowsky, D., Shreevats, A., Gupta, M.: Classifying latent user attributes in twitter. In: Proceedings of the 2nd International Workshop on Search and Mining User-Generated Contents, pp. 37–44. ACM (2010)
11. Burger, J.D., Henderson, J., Kim, G., Zarrella, G.: Discriminating gender on twitter. In: Proceedings of the Conference on Empirical Methods in Natural Language Processing, pp. 1301–1309. Association for Computational Linguistics (2011)
12. Daniel, M. A. Zelenkov, Y.: Russian national corpus as a playground for sociolinguistic research. episode iv. gender and length of the utterance. In: Proceedings of Dialog-2012, pp. 51–62 (2012)
13. Mukherjee, A., Liu, B.: Improving gender classification of blog authors. In: Proceedings of the 2010 Conference on Empirical Methods in Natural Language Processing, pp. 207–217. Association for Computational Linguistics (2010)
14. Vapnik, V.: The nature of statistical learning theory. Data Min. Knowl. Discovery **6**, 1–47 (1995)

15. Al Zamal, F., Liu, W., Ruths, D.: Homophily and latent attribute inference: Inferring latent attributes of twitter users from neighbors. In: ICWSM (2012)
16. Liu, W., Zamal, F.A., Ruths, D.: Using social media to infer gender composition of commuter populations. In: Proceedings of the When the City Meets the Citizen Worksop (2012)
17. Bishop, C.M., Nasrabadi, N.M.: Pattern Recognition and Machine Learning, vol. 1. Springer, New York (2006)
18. Bird, S.: Nltk: the natural language toolkit. In: Proceedings of the COLING/ACL on Interactive Presentation Sessions, pp. 69–72. Association for Computational Linguistics (2006)
19. Agresti, A.: Categorical Data Analysis, vol. 359. Wiley, New York (2002)
20. Panchenko, A., Beaufort, R., Naets, H., Fairon, C.: Towards detection of child sexual abuse media: categorization of the associated filenames. In: Serdyukov, P., Braslavski, P., Kuznetsov, S.O., Kamps, J., Rüger, S., Agichtein, E., Segalovich, I., Yilmaz, E. (eds.) ECIR 2013. LNCS, vol. 7814, pp. 776–779. Springer, Heidelberg (2013)
21. Fan, R.E., Chang, K.W., Hsieh, C.J., Wang, X.R., Lin, C.J.: Liblinear: A library for large linear classification. J. Mach. Learn. Res. **9**, 1871–1874 (2008)
22. Yu, H.F., Huang, F.L., Lin, C.J.: Dual coordinate descent methods for logistic regression and maximum entropy models. Mach. Learn. **85**(1–2), 41–75 (2011)

Clustering Narrow-Domain Short Texts Using K-Means, Linguistic Patterns and LSI

Svetlana Popova[1,2](✉), Vera Danilova[3], and Artem Egorov[2]

[1] Saint-Petersburg State University, Saint Petersburg, Russia
svp@list.ru
http://spbu.ru
[2] ITMO University, Saint-Petersburg, Russia
http://www.ifmo.ru/
[3] Autonomous University of Barcelona, Barcelona, Spain
maolve@gmail.com
http://www.uab.cat/letras/

Abstract. In the present work we consider the problem of narrow-domain clustering of short texts, such as academic abstracts. Our main objective is to check whether it is possible to improve the quality of k-means algorithm expanding the feature space by adding a dictionary of word groups that were selected from texts on the basis of a fixed set of patterns. Also, we check the possibility to increase the quality of clustering by mapping the feature spaces to a semantic space with a lower dimensionality using Latent Semantic Indexing (LSI). The results allow us to assume that the aforementioned modifications are feasible in practical terms as compared to the use of k-means in the feature space defined only by the main dictionary of the corpus.

Keywords: Clustering · Short texts · Narrow domain texts · LSI · Linguistic patterns

1 Introduction

The task of short-text processing is important due to the increase of Internet content, such as news summaries, abstracts, forum messages, social networks, twitter etc. Clustering allows to obtain a structured representation of collections with automatic grouping of topically close documents.

We are interested in clustering academic papers. The solution to this task is required for structured representation of data within scientific domains, e.g. in academic search engines [1–4].

Abstracts are accessed from e-libraries. Many of these libraries usually provide free access to abstracts, while full papers require subscription or a payment.

© Springer International Publishing Switzerland 2014
D.I. Ignatov et al. (Eds.): AIST 2014, CCIS 436, pp. 183–189, 2014.
DOI: 10.1007/978-3-319-12580-0_18

However, abstracts are brief summaries of the contents and are deemed sufficient for clustering academic papers with adequate results [5].

The task of clustering abstracts is closely related to a number of problems, e.g., the task of identifying whether short texts and processed texts may belong to the same topic, which requires the use of an approach to narrow-domain short text clustering [6–10]. One of the main problems with this type of clustering is high data sparseness [8]. If the documents come from the same source it is probable that all of them are topically close. In this case there may be a significant overlap of common words, which complicates the clustering task even more [8]. The corpus for testing and analysis of algorithm performance includes a set of collections that are widely used in the field (CICling, SEPLN-CICling, EasyAbstracts) [5–10]. Experiments implemented within the framework of the present work are also based on these collections.

K-means has been chosen for testing. In the related work, k-means is considered the base or one of the base algorithms for comparison [6,7]. We have not found any papers describing the attempts to improve the quality of k-means performance in narrow-domain short-text clustering, researchers tend to use the base version. In [6,7] it was shown that considering document clustering as an optimization problem ensures high performance. These algorithms yield better results as compared to the base version of k-means. In [12] the advantage of combining LSI with the optimization algorithm is presented, which shows the benefit from the use of LSI in short text clustering (together with algorithms based on particle swarm optimization [6]). We study how to increase the performance of k-means by applying LSI and also by extending the feature space with word groups. The word groups extraction uses linguistic patterns (e.g., NN_NN, NN_NN_NN, JJ_NN, NNS_NN, NN_NNS, JJ_NN_NN, etc., where NN denotes Noun, singular o mass, NNS - Noun, plural, JJ - Adjective; examples of extracted phrases: cluster algorithm, binary vector, maximum entropy model method, etc. that were built on the basis of collections labeled for key phrase extraction tasks.

2   Objectives and Data

2.1   Research Tasks

For the purposes of the present study several tasks were formulated. Firstly, we check whether it is possible to improve the quality of narrow-domain short-text clustering using k-means together with LSI [12]. Secondly, we check the possibility to increase the quality of k-means algorithm performance by using patterns. Thirdly, whether it is possible to improve the clustering quality through the application of both patterns and LSI.

2.2   Testing Dataset

The test set was formed from three collections[1] that are often used for testing narrow-domain short text clustering algorithms. All of them are sets of abstracts

---

[1] http://sites.google.com/site/merrecalde/resources

divided between four clusters by the expert. For each collection the "gold standard" or the best variant of grouping is known. Each collection includes 48 documents. CICling 2002 is considered one of the most difficult collections for clustering [6–8].

2.3 Pattern Extraction

For the purposes of pattern extraction we employed one of the most widely used collections in the field of keyword extraction - SemEval 2010, which was used in the competition of algorithms for keyword extraction TREC 2010 [13]. The collection includes documents and keywords defined by the expert characterizing each document. We used 23 patterns based on the most frequent keywords determined by the expert. We relied on the assumption that, in this way, we will be able to select patterns that are specific to the keywords in texts and also word sequences that reflect textual semantics. We assumed that the use of such sequences would increase clustering quality.

2.4 Clustering

Data Pre-processing. Word stems were used for text representation during clustering. Stemming was performed with Porter Stemmer[2]. Stop-words were removed using a standard list. PoS tags were assigned to each word for the pattern-based extraction of phrases using the Stanford PoS tagger.[3] **Clustering Algorithm.** K-means algorithm was chosen for testing in the present work. Each document is represented as a vector in the feature space defined by the main dictionary or within the feature space formed by the extended dictionary, which includes word groups extracted using patterns. The word order was taken into account during the retrieval of word groups. Upon the extraction each word group was transformed into another group (a set of words). The calculation of term/word group weight was performed using TF-IDF, the distance was measured by the cosine similarity between feature vectors. The number of clusters was the same as in the gold standard, which means that there were 4 clusters in each case. LSI maps the feature space to a semantic space of a lower dimensionality using singular value decomposition of the text-attributes matrix. As a result, each document has a vector representation in the semantic space. To perform LSI we need to indicate the dimensionality of the space where the data are mapped.

---

[2] http://tartarus.org/martin/PorterStemmer/
[3] http://nlp.stanford.edu/software/tagger.shtml

2.5   Evaluation

Clustering quality evaluation was performed in a classic way using the combined information on Precision and Recall of the resulting clusters [14, 15]:

$$F = \sum_i \frac{G_i}{|D|} \max_j F1_{ij}, where\ F1_{ij} = 2 \times \frac{Precision_{ij}\,Recall_{ij}}{Precision_{ij} + Recall_{ij}},$$

$$Precision_{ij} = \frac{|G_i \cap C_j|}{|G_i|}, Recall_{ij} = \frac{|G_i \cap C_j|}{|C_j|}$$

$G = \{G_i\}_{i=\overline{1,m}}$ - clusters generated by the algorithm, $C = \{C_j\}_{j=\overline{1,n}}$ - clusters identified by the experts, D - number of documents in the collection.

3   Experiments and Results

In the course of the experiments two feature space settings were compared: the first was based on the main dictionary of the collection (main dict.) and the second - on the combination of the main dictionary and that of word groups obtained through pattern-based extraction from texts (ext.dict.). For each of the variants we compared the performance of k-means before and after applying the LSI. For each experiment there were 500 iterations on the basis of which the best (max), the worst (min) and the average (avg) results were selected. The results are presented in the Table 1. The values of semantic space dimensionality ("num") that yield the best results are shown.

The analysis of Table 1 allows us to put forward the following assumptions. Firstly, in case of deploying k-means algorithm based on the extended dictionary, no improvement (or a small one) in clustering quality is observed.

Table 1. K-means performance before and after applying LSI ("num" stands for semantic space dimensionality value; "main. dict." indicates all cases where the feature space defined by the main dictionary of the collection was used; "ext. dict." indicates cases where an extended feature space was applied)

	CICling			SEPLN-CICling			EasyAbstracts		
	avg	min	max	avg	min	max	avg	min	max
	with LSI								
Main dict.	0.48	0.35	0.65	0.58	0.36	0.79	0.58	0.36	0.81
Ext. dict	0.49	0.35	**0.66**	0.57	0.35	0.81	**0.59**	0.35	0.86
	with LSI								
Main dict.	**0.53**	**0.48**	0.56	**0.75**	**0.58**	**0.84**	0.51	0.42	0.60
Num (main dict.)	6			3			6		
Ext. dict.	0.49	0.42	0.53	0.74	0.53	0.80	0.49	**0.43**	0.61
Num (Ext. dict.)	7			4			12		

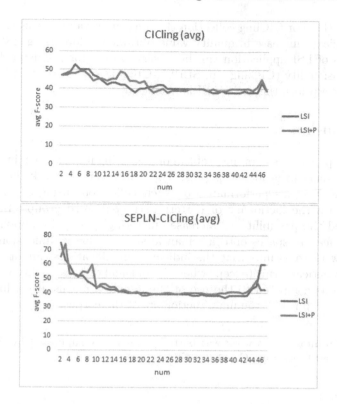

Fig. 1. Dependency of k-means performance on the dimensionality of the semantic space created using LSI (CICling and SEPLN-CICling)

The application of LSI to this feature space does not produce better results than when we use the main dictionary alone. It turns out that in case word groups are extracted using patterns, the results contain much noise (common phrases, such as "experiment results").

Secondly, if the feature space is not expanded with word groups, its mapping to the semantic space with the optimal dimensionality value may lead to some increase in clustering quality. In order to check the feasibility of searching the necessary dimensionality value we modeled the dynamics of conditional dependency between the quality of k-means performance and the dimensionality value.

Figure 1 presents the dependency of the average quality of k-means clustering, according to the results of 500 iterations, on the dimensionality of the semantic space created using LSI (for CICling and SEPLN-CICling). Notation: LSI, if the feature space is based on the main dictionary alone; LSI+P, if the dictionary of word groups is added. According to the diagrams, applying LSI representation, the best result on the narrow-domain short texts can be obtained if the dimensionality of the resulting semantic space is less than 10 (num < 10). Also, the diagrams show that when applying LSI most results are in the range from

0.40 to 0.50 (F). For CICling collection the maximum value of this range causes only insignificant increase in quality value obtained before using LSI. The real effectiveness of LSI application can be observed only for pre-defined values of space dimensionality (CICling - 6, SEPLN-CICling - 3). If these values cannot be identified a priori, the use of LSI is not feasible.

4 Conclusion

The present paper considers the problem of clustering narrow-domain short texts such as abstracts to academic papers. The purpose is to check the possibility of improving k-means performance on such collections using a feature space expanded with the dictionary of pattern-extracted word groups. In addition, we examined the possibility to increase clustering quality by applying LSI to project the feature spaces onto a semantic space of lower dimensionality. The results allow us to assume that the indicated modifications cannot be deemed feasible in practical terms (except when the optimal dimensionality value can be determined) as compared to the use of the simple k-means algorithm and the feature space defined by the main dictionary of the corpus.

Acknowledgement. This work was partially financially supported by the Government of Russian Federation, Grant 074-U01.

References

1. Bernardini, A., Carpineto, C.: Full-subtopic retrieval with keyphrase-based search results clustering. In: IEEE/WIC/ACM International Joint Conferences on Web Intelligence and Intelligent Agent Technologies, vol. 1 (2009)
2. Zhang, D., Dong, Y.: Semantic, hierarchical, online clustering of Web search results. In: Yu, J.X., Lin, X., Lu, H., Zhang, Y. (eds.) APWeb 2004. LNCS, vol. 3007, pp. 69–78. Springer, Heidelberg (2004)
3. Zeng, HJ., He, QC., Chen, Zh., Ma, WY., Ma, J.: Learning to cluster web search results. In: Proceeding SIGIR '04 Proceedings of the 27th Annual International ACM SIGIR Conference on Research and Development in Information Retrieval, pp. 210–217 (2004)
4. Popova, S., Khodyrev, I., Egorov, A., Logvin, S., Gulyaev, S., Karpova, M., Mouromtsev, D.: Sci-search: academic search and analysis system based on keyphrases. In: Klinov, P., Mouromtsev, D. (eds.) KESW 2013. CCIS, vol. 394, pp. 281–288. Springer, Heidelberg (2013)
5. Alexandrov, M., Gelbukh, A., Rosso, P.: An approach to clustering abstracts. In: Montoyo, A., Muñoz, R., Métais, E. (eds.) NLDB 2005. LNCS, vol. 3513, pp. 275–285. Springer, Heidelberg (2005)
6. Cagnina, L., Errecalde, M., Ingaramo, D., Rosso, P.: A discrete particle swarm optimizer for clustering short text corpora. In: BIOMA08, p. 93103 (2008)
7. Errecalde, M., Ingaramo, D., Rosso, P.: ITSA: an effective iterative method for short-text clustering tasks. In: García-Pedrajas, N., Herrera, F., Fyfe, C., Benítez, J.M., Ali, M. (eds.) IEA/AIE 2010. LNCS, vol. 6096, pp. 550–559. Springer, Heidelberg (2010)

8. Pinto, D.: Analysis of narrow-domain short texts clustering. In: Research report for Diploma de Estudios Avanzados (DEA), Department of Information Systems and Computation, UPV (2007)
9. Pinto, D., Rosso, P., Jiménez, H.: A self-enriching methodology for clustering narrow domain short texts. Comput. J. **54**(7), 1148–1165 (2011)
10. Pinto, D., Jiménez-Salazar, H., Rosso, P.: Clustering abstracts of scientific texts using the transition point technique. In: Gelbukh, A. (ed.) CICLing 2006. LNCS, vol. 3878, pp. 536–546. Springer, Heidelberg (2006)
11. Hasanzadeh, E., Poyan, M., Rokny, H.: Text clustering on latent semantic indexing with particle swarm optimization (PSO) algorithm. Int. J. Phys. Sci. **7**(1), 116–120 (2012)
12. Manning, C., Raghavan, P., Schutze, H.: Introduction to Information Retrieval. Cambridge University Press, Cambridge (2009)
13. Kim, S.N., Medelyan, O., Kan, M.Y., Baldwin, T.: Automatic keyphrase extraction from scientific articles. Lang. Resour. Eval. **47**(3), 723–742 (2012)
14. Eissen, S.M., Stein, B.: Analysis of clustering algorithms for Web-based search. In: Karagiannis, D., Reimer, U. (eds.) PAKM 2002. LNCS (LNAI), vol. 2569, pp. 168–178. Springer, Heidelberg (2002)
15. Stein, B., Meyer zu Eissen, S., Wißbrock, F.: On cluster validity and the information need of users. In: Hanza, MH. (ed.) 3rd IASTED International Conference on Artificial Intelligence and Applications (AIA 03), Benalmádena, Spain, pp. 216–221, ISBN 0-88986-390-3. ACTA Press, IASTED (2003)

Analysis of Twitter Users' Mood for Prediction of Gold and Silver Prices in the Stock Market

Alexander Porshnev[✉] and Ilya Redkin

National Research University Higher School of Economics,
Nizhny Novgorod, Russia
aporshnev@hse.ru, ilya-redkin@yandex.ru

Abstract. The question about possibilities to use Twitter users' moods to increase accuracy of stock price movement prediction draws attention of many researchers. In this paper we examine the possibility of analyzing Twitter users' mood to improve accuracy of predictions for Gold and Silver stock market prices. We used a lexicon-based approach to categorize the mood of users expressed in Twitter posts and to analyze 755 million tweets downloaded from February 13, 2013 to September 29, 2013. As forecasting technique, we select Support Vector Machines (SVM), which have shown the best performance. Results of SVM application to prediction the stock market prices for Gold and Silver are discussed.

Keywords: Prediction · Gold · Silver · Twitter · Mood · Psychological states · Support vector machines · Behavioral finance

1 Introduction

Moods, emotions and psychological states of people in situations of uncertainty can influence decision making [1–3]. Stock market prices are usually associated with a high level of uncertainty and a trader's decision can demonstrate a wide set of human cognitive biases and influence of emotional factors. For example, Hirshleifer and Shumway, presuming that sunshine can affect mood, found sunshine to be strongly correlated with stock returns [4].

Over the last years, machine learning and natural language processing have provided tools to use social networks as sources of information. News can be parsed to predict stock price movements. A reaction of the stock market in particular Twitter posts demonstrates that this technique is widely used in common trading practice [5, 6]. Also, the development of linguistic technologies allows a sentiment analysis of Twitter posts to be performed to retrieve moods and psychological states of people. In their pioneer work Bollen and his colleagues reported that analyzing the text content of daily Twitter feeds helped increase the accuracy of DJIA predictions up to 87.6 % [7]. In our preliminary studies we used this approach to DJIA and S&P500 and found out that additional information from Twitter could increase prediction accuracy [8]. Although, works in this area could not provide enough evidence that analysis based on mood and psychological states can provide more reliable and stable information than a news-relied approach that is more sensitive to fake or misleading messages. For example,

© Springer International Publishing Switzerland 2014
D.I. Ignatov et al. (Eds.): AIST 2014, CCIS 436, pp. 190–197, 2014.
DOI: 10.1007/978-3-319-12580-0_19

first Hedge Fund founded by pioneers in Twitter Sentiment Analysis – Derwent Capital Markets' Absolute Return fund failed to demonstrate efficacy and now trying to offer its social media indicators to day traders [9].

In order to receive more information we attempt to test ability of sentiment analysis to predict stock prices. Followed suggestion of the practitioners dealing with different stock market indices we apply Twitter data to the prediction of Gold and Silver stock prices.

Assuming that gold prices can be more influenced by human emotions and psychological states, we expect our data to provide higher accuracy in prediction of gold. Being more related to industry and production by their nature, silver prices are less dependent on moods and emotions. Prediction accuracy for silver prices, therefore, would remain the same even if Twitter information were added.

1.1   Social Mood, Emotions and Decision Making Under Uncertainty

Mood, emotions and decision making are closely connected. In psychological experiments modeling decisions Johnson and Tversky report that psychological states invoked by reading stories can affect the evaluation of risk level [10]. Good moods lead individuals to make more optimistic choices and, vice versa, bad moods lead to pessimistic choices [2, 11, 12]. Good moods and bad moods influence decision making process by invoking the different heuristics, for example individuals in good mood are more likely to eliminate alternatives that did not meet a criterion on an important dimension, that lead to increasing efficiency [13].

The previous section illustrates that an individual's mood affects the decision making process. However, we should not forget that other people play a big role in decision making and shape individual opinions and beliefs. Nofsinger expresses the idea that the general level of optimism/pessimism in society can be connected with economic activity [14]. Nofsinger also supposes that the stock market itself can be a direct measure of social mood [14]. Following Nofsinger, we will regard the economy not as a physical system, but as a complex system of human interactions, in which moods and irrationalities can play a significant role. This point can be supported by observing the informational cascades phenomenon in the stock market [15].

Regarding the stock market and Twitter as two possible measures of social mood, we can assume their correlation and the possibility of using analyses of moods expressed in tweets to increase prediction accuracy for stock market indicators.

2   Methodology

In our research, we use the simple algorithm based on calculation of amount of words from the set of dictionaries (for each emotional category we have the set of words) [8]. The developed twitter sentiment algorithm provides information on a category of emotions a tweet falls within (one or several, we presume that tweets can be multimodal). For example, tweet: *"so happy that brooklyn got into the team."* contain one word "happy" from "Happy" category of our dictionary and this increase the amount of words counted in this category.

Although, proposed approach was really simple and did not took into account possible ambiguity, the F-measure for all the categories was higher than 63 %, we can conclude that it demonstrated a sufficient level of accuracy [16]. The achieved level of classification accuracy helped us obtain a fast and reliable algorithm for sentiment analysis based on the lexicon approach [8].

Algorithm 1. Simple Sentiment Calculation.

Data: Preprocessed Twitter Data

Result: Output: Sum[happy, loving, calm, energetic, fearful, angry, tired, sad]

Find the Dictionaries

foreach Category=[happy, loving, calm, energetic, fearful, angry, tired, sad]

    foreach word in AllTweetsInADay

    If word **In** Dictionaries[Category] **then**

      Sum[Category] + =1

  end

end

return Sum

2.1   Machine Learning in Stock Market Price Forecasts

To test our main hypothesis, we used the Support Vector Machine and Neural Networks to classify days by shifts in stock market indices and use the created model for prediction. The Support Vector Machine and Neural Networks were chosen as most common techniques with the best performance in the field.

In order to answer the question "Does sentiment analysis of tweets provide any additional information?", we used the machine learning techniques on three datasets. The first dataset characterized the stock market over three previous days and was termed the basic set (Basic). The second set was created by adding to the basic set a normalized number of tweets with the words "Worry", "Hope", "Fear" (Basic&WHF). The third set was formed by adding a normalized number of tweets from each of 8 categories of the following emotions: "happy", "loving", "calm", "energetic", "fearful", "angry", "tired", and "sad" (Basic&8EMO). We expect that the comparison between prediction accuracies based on our three learning sets will be different. According to our hypothesis about additional information available in Twitter, we expect the first set to provide the lowest accuracy level, the second set to provide a somewhat higher accuracy, while the highest prediction accuracy will be achieved by using the Basic&8EMO dataset.

In their work, Bollen and his co-authors found better predictions based on data of four previous days, and adding data about more days led to an overtraining model [7]. To test these findings, we trained the Support Vector Machine and Neural Networks on data from Twitter with different time lags (from one to seven days).

2.2 Data Description

By making use of Twitter API we managed to download more than 700 million tweets over the period from 13/02/2013 to 29/09/2013 (we downloaded an average of 3,483,642 tweets per day). All the tweets were sorted by day, analyzed automatically according to data counts of the words "Worry", "Hope", "Fear" (WHF dataset) and assigned by the developed sentiment analyzer counting tweets in the following categories: "happy", "loving", "calm", "energetic", "fearful", "angry", "tired", and "sad" (8EMO dataset).

For the Gold and Silver stock market prices data (GCG4, SIH4) we used the InvestorPoint website (http://www.investorpoint.com), which provides opening and closing historical prices as well as the volume for any given trading day. To apply the machine learning techniques, we divided the days into two groups by adding a variable growth. Variable growth equals to 1, when the opening price was lower than the price at close or equals 0, when the opening price was higher than or equal to the price at close. As a result, the Basic dataset consisted of 16 columns (variables: growth, $Open_{-1 \ day}$, $Close_{-1 \ day}$, $Min_{-1 \ day}$, $Max_{-1 \ day}$, $Volume_{-1 \ day}$, $Open_{-2 \ day}$, $Close_{-2 \ day}$, $Min_{-2 \ day}$, $Max_{-2 \ day}$, $Volume_{-2 \ day}$; $Open_{-3 \ day}$, $Close_{-3 \ day}$, $Min_{-3 \ day}$, $Max_{-3 \ day}$, $Volume_{-3 \ day}$).

The first column provided information about the index shift (1 or 0), then presented the opening price, closing price, maximum price, minimum price and volume for three previous days. The Basic&WHF dataset was created by adding columns about frequencies of the words "worry", "hope" and "fear" for the previous day (one day – three columns) or for several days (3 × number of days). For example, the Basic&WHF dataset for the previous 7 days consisted of 37 columns (16 + 3 × 7). While the Basic&8EMO dataset is formed in the same way, the three columns with word frequencies are replaced by 8 columns with frequencies of the words from the developed dictionary of emotional states. For example, the Basic&8EMO set with data from the sentiment analysis of the previous 7 days is composed of 72 columns (16 + 8 × 7).

The whole period from 13/02/2013 to 29/09/2013 was divided into sets with data of 120 days. The period of 120 days was chosen to enable the use of 105 days for training and 15 days for prediction. Within that period we ran a minimum of 5 experiments with the dataset containing information with a lag from one to seven days (75 predictions) and a maximum of 10 experiments (150 predictions), depending on the availability of stock market data.

3 Analysis

3.1 Gold Prices Stock Market Prediction

First we tried to find the baseline accuracy and applied Neural Networks and Support Vector Machine trained only on the Basic DJIA data with time lags from one to seven days (Table 1). The best accuracy was demonstrated by the Support Vector Machine with a time lag of two days (70 %). That level of accuracy became baseline for our analysis. The Support Vector Machine provided better performance and we used this technique for further analysis.

Table 1. Prediction of growth or fall for Gold prices. Accuracy of the Support Vector Machines and Neural Networks trained on the Basic dataset

Lag in days	1	2	3	4	5	6	7
SVM	69.52 %	**70.00 %**	67.62 %	66.67 %	69.33 %	64.67 %	60.67 %
Neural Net	52.38 %	54.00 %	57.14 %	54.29 %	66.00 %	62.67 %	56.67 %
Days for prediction	150	150	150	150	150	150	150

The next Support Vector Machine was trained on extended datasets with Twitter information (Table 2). The accuracies presented in Table 2 show that extended datasets lead to model overtraining and impair prediction accuracy in comparison with 70 % gained with just historical data applied.

Table 2. Prediction of growth or fall for Gold prices. Accuracy of the Support Vector Machines and Neural Networks trained on the Basic dataset

Dataset	Number of previous days included in dataset						
	1	2	3	4	5	6	7
Basic&WHF	37.14 %	40.00 %	43.81 %	43.81 %	48.00 %	55.33 %	**64.67 %**
Basic&8EMO	42.86 %	37.33 %	40.95 %	42.86 %	44.00 %	46.67 %	55.33 %
Number of experiments	7	10	7	7	10	10	10
Number of days for prediction	105	150	105	105	150	150	150

These results reject our hypothesis that Twitter data could be used to increase prediction accuracy for Gold stock market prices.

3.2   Silver Prices Stock Market Prediction

To establish the baseline accuracy, we ran Neural Networks and the Support Vector Machine on historical data (Table 3). The results showed that the Support Vector Machine provided a better accuracy (72.50 %) than Neural Networks (69.33 %). The best performance was observed on data with a 6 days lag.

The added Twitter information failed to improve prediction accuracy (Table 4). The best performance achieved by adding counts of the words "worry", "hope", and "fear" was 71.65 %, which is lower than 72.50 % obtained using just historical information.

Analysis of correlations. As we found in the previous section, the application of machine learning techniques failed to increase forecast accuracy, but keeping in mind our main idea of the stock market and Twitter expressing social mood, we decided to

Table 3. Prediction of growth or fall for Silver prices. Accuracy of the Support Vector Machines and Neural Networks trained on the Basic dataset

Lag in days	1	2	3	4	5	6	7
SVM	61.33 %	33.33 %	58.33 %	42.67 %	61.11 %	**72.50 %**	37.78 %
Neural Net	69.33 %	55.00 %	55.00 %	46.67 %	62.22 %	53.33 %	41.11 %
Number of days for prediction	75	75	75	75	75	75	75

Table 4. Stock market prices for Silver prediction Accuracy of the Support Vector Machine versus the training dataset

Dataset	Number of previous days included in dataset						
	1 day	2 days	3 days	4 days	5 days	6 days	7 days
Basic&WHF	41.33 %	42.22 %	48.33 %	45.33 %	50.00 %	**71.67 %**	42.22 %
Basic&8EMO	42.67 %	42.22 %	58.33 %	44.00 %	44.44 %	69.17 %	43.33 %
Number of experiments	5	6	4	5	6	8	6
Number of days for prediction	75	90	60	75	90	120	90

analyze the correlation between the normalized amount of tweets in each category and silver and gold prices for in the stock market (Table 5). According to our data format, we took one file with 120 days without a time lag. The results were surprising: Open Price appeared to be in a highly positive correlation with almost all the emotional scales from Twitter, but it was more closely correlated with Max Price and Min Price. We can see the same results for Gold and Silver stock market prices on all 8 files with 120 days intervals.

It is evident that the prices for this period are so highly correlated that addition of information from Twitter will not provide more information. However, the average

Table 5. Pearson's correlation of Gold and Silver Open Prices and others variables presented in the dataset (120 days, 19.02.2013–18.09.2013).

Variables presented in dataset Basic&8EMO	Silver	Gold
Max price	0.995**	0.997**
Min price	0.996**	0.994**
Close price	−0.553**	−0.365**
Happy	0.265*	0.405**
Loving	0.452**	0.413**
Calm	0.204*	0.262*
Energetic	0.655**	0.650**
Fearful	0.019	0.062
Angry	0.574**	0.561**
Tired	0.778**	0.752**

* Correlation is significant at the 0.01 level, ** Correlation is significant at the 0.05 level

correlation between stock prices and emotions (with categories $r_{\text{Tired}} = 0.778$ and $r_{\text{Energetic}} = 0.655$) expressed in Twitter supports the social mood idea.

4 Discussion

The application of Twitter data to stock market prediction looks like an attempt to use a magic crystal ball or unrelated data. However, analyzing the possibility of improving prediction accuracy for stock market indices (DJIA, S&P500 and NASDAQ), we found that Twitter sentiment analysis data could significantly improve forecasting for the S&P500 index. Addition of Twitter information allows us to increase accuracy from 57 % (baseline) to 70 % (SVM trained on Basic&8EMO data). We discovered this effect by running 17 experiments within 15 days for predictions, 255 days in total.

Although this study lacks the same results, we did not achieve any significantly higher accuracy for both Silver and Gold stock market prices. We think this can be explained by the high level of prediction provided by the baseline data of 70 % (Gold) and 72.50 % (Silver). Addition of Twitter information may result in overfitted models and will not provide a better accuracy.

Comparison of the results obtained on two datasets with added Twitter data showed that the technique trained on Basic&WHF performed no worse than that trained on Basic&8EMO. We observe the same results in our work devoted to analysis of stock market indices. There are two possible explanations: first, Basic&8EMO provided more data that led to model overtraining; the second explanation is connected with the need to further improve sentiment analysis of tweets. Enhanced accuracy of sentiment analysis may deliver better results Potential areas of improvement may be introducing the weight of words, for example, the word "fear" should be heavier in tweet analysis than, for example, "coward".

5 Conclusions

In our research we planned to test the hypothesis that sentiment analysis of Twitter data could provide additional information that may improve the prediction accuracy for Gold and Silver prices, as stock market traders usually regard them as a more mood dependent. We developed a server application to download, store and analyze tweets. Over a period from 13/02/2013 to 29/09/2013, we downloaded 755,000,101 tweets. Comparison of SVM application on historical data and data from Twitter showed that at this market our sentiment analysis could not improve accuracy of forecast. The obtained results suggest that our hypothesis cannot be confirmed. Addition of Twitter information failed to help us make a more accurate forecast. However, we found this set of stock market prices to be highly auto correlated, which may provide the basis for a successful trading strategy. In our further research we plan to enhance accuracy by improving dictionaries and introducing word weights, thoroughly check the reliability and validity of our dictionaries' algorithm continue data collection, and implement machine learning techniques to predict the amount of change for chosen stock market indices.

References

1. Ding, T., Fang, V., Zuo, D.: Stock market prediction based on time series data and market sentiment (2013). http://murphy.wot.eecs.northwestern.edu/~pzu918/EECS349/final_dZuo_tDing_vFang.pdf. Accessed 30 Jun 2013
2. Mayer, J.D., Gaschke, Y.N., Braverman, D.L., Evans, T.W.: Mood-congruent judgment is a general effect. J. Pers. Soc. Psychol. **63**, 119 (1992)
3. McFarland, C., White, K., Newth, S.: Mood acknowledgment and correction for the mood-congruency bias in social judgment. J. Exp. Soc. Psychol. **39**, 483–491 (2003). doi:10.1016/S0022-1031(03)00025-8
4. Hirshleifer, D., Shumway, T.: Good day sunshine: Stock returns and the weather. J Finance **58**, 1009–1032 (2003)
5. Prezioso, J.: Yom Kippur war tweet prompts higher oil prices. In: Reuters (2013). http://www.huffingtonpost.com/2013/10/10/yom-kippur-war-tweet-oil-prices-traders_n_4079634.html. Accessed 22 Jan 2014
6. Selyukh, A.: Hackers send fake market-moving AP tweet on White House explosions. In: Reuters (2013). http://www.reuters.com/article/2013/04/23/net-us-usa-whitehouse-ap-idUSBRE93M12Y20130423. Accessed 17 Sep 2013
7. Bollen, J., Mao, H., Zeng, X.: Twitter mood predicts the stock market. J. Comput. Sci. **2**, 1–8 (2011). doi:10.1016/j.jocs.2010.12.007
8. Porshnev, A., Redkin, I., Shevchenko, A.: Improving Prediction of Stock Market Indices by Analyzing the Psychological States of Twitter Users. Social Science Research Network, Rochester (2013)
9. Mackintosh, J., Editor, I.: Last tweet for Derwent's Absolute Return. Financial Times (2012)
10. Johnson, E.J., Tversky, A.: Affect, generalization, and the perception of risk. J. Pers. Soc. Psychol. **45**, 20 (1983)
11. Isen, A.M., Patrick, R.: The effect of positive feelings on risk taking: When the chips are down. Organ. Behav. Hum. Perform. **31**, 194–202 (1983)
12. Schwarz, N., Clore, G.L.: Mood, misattribution, and judgments of well-being: Informative and directive functions of affective states. J. Pers. Soc. Psychol. **45**, 513 (1983)
13. Isen, A.M., Means, B.: The influence of positive affect on decision-making strategy. Soc. Cogn. **2**, 18–31 (1983)
14. Nofsinger, J.R.: Social mood and financial economics. J. Behav. Finance **6**, 144–160 (2005). doi:10.1207/s15427579jpfm0603_4
15. Bikhchandani, S., Hirshleifer, D., Welch, I.: A theory of fads, fashion, custom, and cultural change as informational cascades. J. Polit. Econ. **100**, 992–1026 (1992)
16. Chen, R., Lazer, M.: Sentiment analysis of twitter feeds for the prediction of stock market movement. In: stanford.edu (2013). http://cs229.stanford.edu/proj2011/ChenLazer-SentimentAnalysisOfTwitterFeedsForThePredictionOfStockMarketMovement.pdf. Accessed 25 Jan 2013

Automatic Contrast Parameter Estimation in Anisotropic Diffusion for Image Restoration

V.B. Surya Prasath[1](✉) and Radhakrishnan Delhibabu[2]

[1] University of Missouri, Columbia, MO 65211, USA
prasaths@missouri.edu
http://web.missouri.edu/~prasaths
[2] Informatik 5, KBSG, RWTH Aachen, Aachen, Germany
rdelhibabu@kbsg.rwth-aachen.de

Abstract. Anisotropic diffusion is used widely in image processing for edge preserving filtering and image smoothing tasks. One of the important class of such model is by Perona and Malik (PM) who used a gradient based diffusion to drive smoothing along edges and not across it. The contrast parameter used in the PM method needs to be carefully chosen to obtain optimal denoising results. Here we consider a local histogram based cumulative distribution approach for selecting this parameter in a data adaptive way so as to avoid manual tuning. We use spatial smoothing based diffusion coefficient along with adaptive contrast parameter estimation for obtaining better edge maps. Moreover, experimental results indicate that this adaptive scheme performs well for a variety of noisy images and comparison results indicate we obtain better peak signal to noise ratio and structural similarity scores with respect to fixed constant parameter values.

Keywords: Image restoration · Anisotropic diffusion · Contrast parameter · Local histogram · Denoising

1 Introduction

Image restoration is one of the classical problems in digital image processing. It has been studied for the last five decades and recent advancements in imaging technologies have made the task of automatically removing noise one of the paramount research problems. Variational and partial differential equation (PDE) based schemes are getting popular due to their edgwe preserving denoising and selective smoothing of images [1].

Perona and Malik [2] (PM for short) initiated the anisotropic diffusion PDE model, which is a second order time-dependent parabolic equation for computing edges and scale space of images. Despite its success in image denoising and edge detection the PM PDE is known to create blocky or staircasing artifacts, [3]. Moreover, the diffusion PDEs involve parameters which needs to be tuned to obtain optimal denoising results. To mitigate this various proposals has been

© Springer International Publishing Switzerland 2014
D.I. Ignatov et al. (Eds.): AIST 2014, CCIS 436, pp. 198–206, 2014.
DOI: 10.1007/978-3-319-12580-0_20

made over the last few years, for example, inverse gradient [3], hybrid edge detectors [4,5] and other ad-hoc modifications.

In this paper, we study an adaptive version of the well-known Perona and Malik PDE by utilizing local histograms information in a dynamic manner. One of the main ingredient for successful denoising result is the contrast parameter which appears in the nonlinear diffusion coefficient. Instead of fixing the contrast parameter manually which can lead to variable results such as excessive blurring or no denoising, we use the local histogram to drive the diffusion. To avoid ill-posedness we make use of the spatial regularization approach of Catte et al. [6]. We provide preliminary experimental results on a variety of noisy test images to show that the proposed method performs better than the traditional PDE based schemes. Compared with manually fixing the contrast parameter and Canny edge detector based suggestion made in [2] our adaptive local histogram approach obtained superior results as well.

The rest of the paper is organized as follows. Section 2 gives the proposed adaptive diffusion based denoising PDE model. Section 3 provides experimental results on various images corrupted by Gaussian noise supporting our proposed approach. Finally, Sect. 4 concludes the paper.

2   Adaptive Diffusion Scheme

2.1   Perona-Malik PDE

Perona and Malik [2] studied the following (continuous setting) PDE for image restoration,

$$\frac{\partial u(x,t)}{\partial t} = \nabla \cdot \Big(g(|\nabla u(x,t)|^2) \nabla u(x,t) \Big) - \mu \left(u(x,t) - u_0(x) \right) \tag{1}$$

where $u_0 : \Omega \subset \mathbb{R}^2 \to \mathbb{R}$ is the given input (noisy) image, $x = (x_1, x_2) \in \Omega$ pixel co-ordinates on the image domain Ω, $\nabla = (\partial_{x_1}, \partial_{x_2})$ is the gradient operator, $\mu \geq 0$ fidelity parameter, wand $g(\cdot)$ is the diffusion coefficient which tunes the amount of smoothing according to the gradient magnitude. Here the artificial temporal variable "t" provides a series of images $\{u(x,t)\}_{t=0}^{\infty}$ which creates a nonlinear scale-space starting with the given initial noisy image $u(x,0) = u_0(x)$. In the discretized version of the PDE (1) the time variable "t" is equivalent to the number of iterations and higher number of iterations induce longer time diffusion thereby obtaining smoother results. If $g \equiv 1$ then we see that the PDE in Eq. (1) is the Heat (diffusion) equation which is known obtain smoother solutions. In their seminar work [2], Perona and Malik advocate decreasing diffusion coefficients so as to reduce the amount of smoothing in regions where the gradients are high (which indicate the possibility of edge pixels) and keep the smoothing of Heat equation in regions where the gradients were low (which are flat/homogenous regions). For example the two decreasing functions,

$$g_1(|\nabla u(x,t)|^2) = \frac{1}{1 + |\nabla u(x,t)|^2/K^2}, \tag{2}$$

$$g_2(|\nabla u(x,t)|^2) = \exp\left(-|\nabla u(x,t)|^2/K^2\right), \tag{3}$$

are prime examples which are used effectively as proposed in the original work of [2]. Here $K > 0$ is a tunable parameter (the so-called *contrast parameter*) and is crucial in obtaining edge preserving restorations. We can see that the PM PDE (1) defines two conflicting regimes of diffusion flows on a given image u_0 as the time progresses [7]:

- Forward diffusion: Inside the regions where the magnitude of the gradient of $u(x,t)$ is weak, i.e., $|\nabla u(x,t)| < K$, the Eq. (1) is the Heat equation which is known to give very smoothing result.
- Backward diffusion: Near the edges where the magnitude of the gradient $u(x,t)$ is large, i.e., $|\nabla u(x,t)| \geq K$, the Eq. (1) resorts to inverse Heat equation which is known to sharpen the edges.

Thus, we see that the contrast parameter $K > 0$ is very important in obtaining meaningful restoration results and is traditionally fixed by tuning manually according to the domain of the images. The PM PDE Eq. (1) along with either one of the diffusion coefficients given in Eqs. (2) and (3) promotes combined forward-backward diffusion flow. Due to this forward-backward flow combination the PM PDE is ill posed and various well-posed alternatives have been considered before [4,6,8,9]. One of the basic model to avoid ill-posedness is to make use spatial smoothing in gradient computations [6] and is obtained by changing the diffusion coefficients argument as follows,

$$g = g(|G_\rho \star \nabla u(x,t)|^2) \tag{4}$$

where G_ρ is the 2D Gaussian kernel with ρ the standard deviation,

$$G_\rho = (\rho\sqrt{2\pi})^{-1} \exp\left(-|\mathbf{x}|^2/2\rho^2\right), \tag{5}$$

and $\star$ denotes the $2D$ convolution operation.

2.2   Adaptive Contrast Parameter

Local histogram are widely utilized in various image processing problems [10]. It is based on local information around a pixel as they are computed from patches and thus can lead to better differentiation of multi-scale features present in digital images. For a given discrete gray-scale image $u : \Omega \subset \mathbb{N}^2 \to [0,255]$, let $W_{x,r}$ be the local neighborhood region centered at a pixel $x \in \Omega$ with radius r. We compute the local histogram of the pixel $x \in \Omega$ and its corresponding cumulative distribution function (CDF),

$$P_x^t(y) = \frac{|\{z \in W_{x,r} \cap \Omega \,|\, u(z,t) = y\}|}{W_{x,r} \cap \Omega}, \tag{6}$$

$$F_x^t(y) = \frac{|\{z \in W_{x,r} \cap \Omega \,|\, u(z,t) \leq y\}|}{W_{x,r} \cap \Omega} \tag{7}$$

Fig. 1. Adaptive contrast parameter (rescaled to $[0,1]$ for visualization purpose) for some standard gray-scale test images. For noise-free (top row) and noisy (Gaussian noise added, $\sigma = 30$) (bottom row) images. As can be seen the local histogram based diffusion function $\tilde{g}(|G_\rho \star \nabla u(x,t)|^2)$ (with $\rho = 5$, see Eq. (9)) captures salient edges even under severe noise.

for $0 \leq y \leq 255$, respectively. Using the CDF we define an adaptive contrast parameter $\xi(x,t)$ for each $x \in \Omega$ at time $t > 0$,

$$\xi(u(x,t-1)) = \frac{\int_0^{255} F_x^{t-1}(y)\,dy}{\max\limits_{x \in \Omega} \int_0^{255} F_x^{t-1}(y)\,dy}. \tag{8}$$

We utilize the local histogram based measure at the previous time $(t-1)$ to dynamically update the current contrast parameter in the diffusion PDE (1) with $g_1(\cdot)$ in (2),

$$\tilde{g}(|G_\rho \star \nabla u(x,t)|^2) = \frac{1}{1 + |G_\rho \star \nabla u(x,t)|^2 / \xi(u(x,t-1))^2} \tag{9}$$

Note that this adaptive parameter depends on each pixel and captures the local statistics around that pixel under consideration, see Fig. 1. Thus, the proposed adaptive PM PDE scheme is given by,

$$\begin{cases} \dfrac{\partial u(x,t)}{\partial t} = \nabla \cdot \left(\tilde{g}(|G_\rho \star \nabla u(x,t)|^2)\, \nabla u(x,t) \right) - \mu(u(x,t) - u_0(x)), & x \in \Omega, \\ u(x,0) = u_0(x), & x \in \Omega, \\ \dfrac{\partial u}{\partial n} = 0, & x \in \partial\Omega. \end{cases} \tag{10}$$

The wellposedness of the above model can be proved following the arguments in [6] and will be reported elsewhere. Next, we provide preliminary experimental results indicating the validity of our proposed adaptive diffusion model (10) on noisy standard test images.

3   Experimental Results

3.1   Setup and Parameters

All the images and contrast parameter values are mapped to $[0,1]$ range in our experiments. We take the fidelity parameter $\mu = 0$ in our experiments to concentrate on the effect of the contrast parameter in the diffusion process (1). The pre-smoothing parameter $\rho = 5$ in the Gaussian kernel (see Eq. (5)) is fixed for the high Gaussian noise level $\sigma \geq 25$, and reduced to $\rho = 2$ for noise level low-medium $5 \leq \sigma \leq 25$. The neighborhood $(W_{x,r})$ size in the contrast parameter estimation is set at $r = 3$, see Eqs. (6)–(8). The discretized version of PM PDE in Eq. (10) using explicit Euler finite difference scheme can be written as follows [11], (with initial $u_{i,j}^0 = (u_0)_{i,j}$)

$$u_{i,j}^{n+1} = u_{i,j}^n + \frac{\Delta t}{|\mathcal{N}_{i,j}|} \sum_{(k,l) \in \mathcal{N}_{i,j}} \tilde{g}(|G_\rho \star \nabla u|_{k,l}^n) \cdot |u_{k,l}^n|, \qquad (11)$$

where $u_{i,j}^{n+1}$ is the discrete image at time $n + 1$, $\mathcal{N}_{i,j}$ is a neighborhood of pixel (i, j), and Δt is the temporal step size which is need to be ≤ 0.25 for stability. The local histogram computations take 6 seconds per iteration whereas solving the discrete diffusion PDE (11) for an iteration takes 0.006 for a gray-scale image of size $|\Omega| = 256 \times 256$ pixels. Note that the experiments were performed on a Mac Pro Laptop with 2.3 GHz Intel Core i7 processor, 8 Gb RAM memory and MATLAB R2012a is used for the calculations and visualizations. We believe optimizing and utilizing C/C++ will decrease the computational times further.

3.2   Comparison Results

Perona and Malik in [2] suggest using a Canny edge detector [12] based estimation for the contrast parameter. Figure 2 provides comparison of the PM PDE (1) with the diffusion coefficient function $g_1(\cdot)$ in (2), for different parameter K values and adaptive Canny and our local histogram based method for the *Cameraman* gray scale 256×256 image. We show the original (Fig. 2(a)) and noisy image (Fig. 2(b)), the high noise level $\sigma = 30$ is chosen to test the robustness and I can be seen by comparing with manually fixed (Fig. 2(c-f)) both the Canny based (Fig. 2(g)) and our approach (Fig. 2(h)) obtain better restoration results without noisy speckles. Moreover, our approach obtains the optimal result as against the Canny based which smooths edges (see chin, mouth regions). This is further supported by the PSNR values given in the Fig. 2, which indicate we get the best possible result. The PSNR (given in dB) is given by the following formula for a gray scale image,

(a) Original

(b) Noisy (18.56)

(c) $K = 0.25$ (26.45)

(d) $K = 0.50$ (25.67)

(e) $K = 0.75$ (24.22)

(f) $K = 1$ (23.02)

(g) Canny (26.82)

(h) Proposed (**27.30**)

Fig. 2. Comparison of restoration results for manually tuned contrast parameter K versus automatic estimations for the PM PDE Eq. (1) with PSNR (dB) value for each is given in the parentheses. (a) Original image of size 265×265 (b) Gaussian noise ($\sigma = 30$) corrupted image (c-f) Manually set fixed K values with diffusion coefficient in Eq. (2) (g) Canny edge detector based estimation for K and (h) Proposed local histogram based adaptive diffusion coefficient (9).

$$\mathrm{PSNR}(u) := 20 * \log_{10}\left(\frac{u_{max}}{\sqrt{\mathrm{MSE}}}\right) dB \tag{12}$$

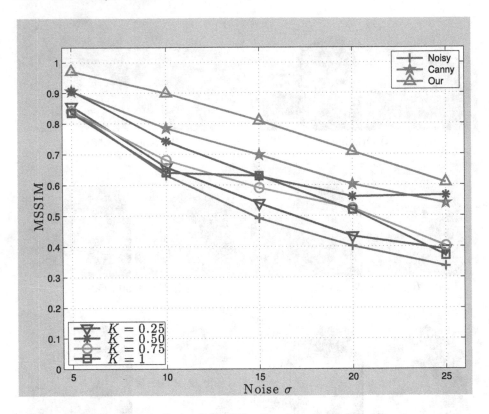

Fig. 3. Mean structural similarity (MSSIM) values for restoring the *Cameraman* image using PM PDE Eq. (1) with diffusion coefficient in Eq. (2) for different constant K values against our adaptive estimation based method.

where MSE $= (mn)^{-1} \sum \sum (u - u_O)^2$, with u_O is the original (noise free) image, $m \times n$ denotes the image size, u_{max} denotes the maximum value, for example in 8-bit images $u_{max} = 255$. A difference of $0.5\,dB$ can be identified visually. Though PSNR provides the amount of signal improvement after restoration under noisy scenario it does not capture structural similarity between noise-free and restored results. Mean structural similarity (MSSIM) metric is known to provide a better measure of restored images [13]. The MSSIM value near one implies the optimal denoising capability of a scheme. In Fig. 3 we show the restoration results on noisy *Cameraman* image corrupted by Gaussian noise of variance elves $\sigma = 5$ to 25 with different K values and adaptive estimation methods. As can be seen we obtain the optimal MSSIM values overall and among other parameter estimations taking $K = 0.50$ performs better in higher noise levels and Canny based method obtains better in low noise levels.

Finally, in Table 1 we show the PSNR and MSSIM results for some of the well-known standard test images with Gaussian noise $\sigma = 30$ for different parameter K values and adaptive methods. As can be seen our local histogram based

Table 1. PSNR (dB)/MSSIM comparison for different standard test images under noise level $\sigma = 30$. Noisy images (18.56/0.3292) were obtained by adding Gaussian noise of strength σ to original images. Best results are in **boldface** and second-best underlined.

Image	K value				Our
names	0.25	0.50	0.75	1	Proposed
Lena	<u>26.20</u>/0.6722	25.36/<u>0.7456</u>	24.14/0.7022	23.17/0.6674	**27.44/0.7860**
Barbara	<u>23.86</u>/<u>0.7970</u>	22.58/0.7326	21.91/0.6888	21.66/0.6677	**24.34/0.8107**
House	<u>28.11</u>/0.7647	27.82/<u>0.8115</u>	26.30/0.7816	24.96/0.7547	**29.82/0.8219**
Baboon	<u>23.89</u>/0.5480	22.97/0.4224	22.57/0.3859	22.43/<u>0.3741</u>	**25.06/0.4646**

approach obtains best results in all. This indicates that our method works well overall and the noise is removed with structure preservation.

4 Conclusion

In this paper we study an adaptive version of the well-known Perona and Malik anisotropic diffusion for image restoration. By utilizing a local histogram based pixel-wise estimation as a contrast parameter we obtain effective denoising without manual tuning of parameters. Experimental results on different noisy images show that the proposed adaptive anisotropic diffusion scheme denoises without excessive smoothing associated with single scalar contrast parameter based results. Moreover, compared with different contrast parameter values adaptive version outperforms all with peak signal to noise ratio gain and structural similarity to noise-free on test images.

References

1. Aubert, G., Kornprobst, P.: Mathematical problems in image processing: Partial differential equation and calculus of variations. Applied Mathematical Sciences, vol. 147, 2nd edn. Springer-Verlag, New York (2006)
2. Perona, P., Malik, J.: Scale-space and edge detection using anisotropic diffusion. IEEE Trans. Pattern Anal. Mach. Intell. **12**(7), 629–639 (1990)
3. Prasath, V.B.S., Singh, A.: A hybrid convex variational model for image restoration. Appl. Math. Comput. **215**(10), 3655–3664 (2010)
4. Prasath, V.B.S.: A well-posed multiscale regularization scheme for digital image denoising. Int. J. Appl. Math. Comput. Sci. **21**(4), 769–777 (2011)
5. Prasath, V.B.S., Singh, A.: An adaptive anisotropic diffusion scheme for image restoration and selective smoothing. Int. J. Appl. Math. Comput. Sci. **12**(1), 1–18 (2012)
6. Catte, V., Lions, P.L., Morel, J.M., Coll, T.: Image selective smoothing and edge detection by nonlinear diffusion. SIAM J. Numer. Anal. **29**(1), 182–193 (1992)
7. Prasath, V.B.S., Singh, A.: Well-posed inhomogeneous nonlinear diffusion scheme for digital image denoising. J. Appl. Math. **2010**, 1–14 (2010). Article ID 763847

8. Prasath, V.B.S., Vorotnikov, D.: Weighted and well-balanced anisotropic diffusion scheme for image denoising and restoration. Nonlinear Anal.: Real World Appl. **17**, 33–46 (2013)
9. Prasath, V.B.S., Vorotnikov, D.: On a system of adaptive coupled PDEs for image restoration. J. Math. Imaging Vis. **48**(1), 35–52 (2014). arXiv:1112.2904
10. Prasath, V.B.S., Palaniappan, K., Seetharaman, G.: Multichannel texture image segmentation using weighted feature fitting based variational active contours. In: Eighth Indian Conference on Vision, Graphics and Image Processing (ICVGIP), Mumbai, India, December 2012
11. Prasath, V.B.S., Moreno, J.C.: On convergent finite difference schemes for varitional - PDE based image processing. Technical report, ArXiv (2013). http://arxiv.org/abs/1310.7443
12. Canny, J.F.: A computational approach to edge detection. IEEE Trans. Pattern Anal. Mach. Intell. **8**(6), 679–698 (1986)
13. Wang, Z., Bovik, A.C., Sheikh, H.R., Simoncelli, E.P.: Image quality assessment: from error visibility to structural similarity. IEEE Trans. Image Process. **13**(4), 600–612 (2004)

Evaluation of the Delta TF-IDF Features for Sentiment Analysis

Andrew B. Samoylov[(⊠)]

Southern Federal University, Rostov-on-Don, Russia
burgundez@nextmail.ru

Abstract. This paper proposes a feature model different from bag-of-word models to analyze the sentiment of the text. The main idea of the method is improving the quality of prediction by combining a rule-based approach and the standard bag-of-words model. Results of the experiments with changing the subject, the size of reviews in data are shown. The hypothesis stating that it is better to use short message with the length of 1–2 sentences or tweets for calculation Delta TFIDF was tested.

Keywords: Sentiment analysis · Tweets · Subjective an objective sentences · Delta TFIDF

1 Introduction

Sentiment analysis is a class of methods of content analysis in computational linguistics, designed for automated detection of emotive vocabulary in texts and authors' attitude to the objects that they referred in the text. The main and the best results of analysis are based on machine learning methods, such as SVM [4–6], Naive Bayes classifier (NB) [6], patterns of emotions [1] and different linguistic rules [6,7]. It should be noted that many of the works are based on paper [2]. That is to say the mathematical model of the text is the so-called bag-of-words, and the main method of prediction is the machine learning method (SVM, NB). This idea was developed in a successive work of the above-mentioned authors [3], by adding the subjectivity analysis of text. In the subsequent years various attempts to improve their results were made. It is necessary to note the works, where the elements of the theory of information retrieval were used, such as term-frequency (TF) and invert document-frequency (IDF) and their analogue for sentiment analysis Delta TFIDF [4,5].

To increase model quality, one should take into account important lexical items such as negation, intensifiers, etc. (list of such lexical structures can be found in [1,2]). The importance of these structures is shown in the following example: phrase "I love the film" consists almost of the same words as "I don't love the film", but they express opposite emotions. Modern online classifiers handle such lexical structures in different ways, but unfortunately not always

© Springer International Publishing Switzerland 2014
D.I. Ignatov et al. (Eds.): AIST 2014, CCIS 436, pp. 207–212, 2014.
DOI: 10.1007/978-3-319-12580-0_21

correctly. For example Stanford classifier[1] makes mistakes when dealing with the treatment of such phrases as: "I don't love the film" and "I have not been disappointed by the film!", classifying them as positive and negative. The NB classifier[2] gives the same result. Thus, one of the main objectives of the article is the development of a method for sentiment analysis, which takes into account different lexical structures.

The next problem, studied in the paper, is choice of a corpus for calculating Delta TFIDF. There are papers where cross domain models are studied [5,7,8]. It is a good way to examine the flexibility of a model. But the influence of the reviews size on trained classifier is investigated insufficiently. To emphasize importance of the problem, consider the following example: "I love Star Wars". "Star Wars" has a positive value here. Now consider the following review: "I don't like films about scientific fantasy. These films are very casual. But I remember some exceptions. Star Track is interesting space story. Star wars too. But, unfortunately, major part of space films is cheap fake". "Star Wars" doesn't have negative value, but due to a generally negative review a classifier would learn to recognize it as negative. It is natural to assume that for the best performance of a classifier, it should use small reviews. The main goal of the paper is to test this hypothesis for Delta TFIDF calculation.

Summarizing the main objectives of this paper are

1. To develop a model that would improve the quality of sentiment analysis.
2. To find out whether the subject of the corpus for calculating Delta TFIDF influences the overall quality of the prediction for designed model.
3. To test the hypothesis that for Delta TFIDF calculation it is better to use examples of text with a small amount of words.

The paper is organized as follows: in the second section the method of text analysis is described. Then in the third section the necessary theoretical background and features used to describe the model are shown. Fourth section is devoted to the experiments and their analysis. The fifth section is the conclusion.

2   The Method of Text Analysis

We denote corpus for calculating weights of words by CW. For sentiment analysis of a text we use the following algorithm.

1. For a given corpus CW calculate values of weight for words.
2. Repeat for each document from train sample:
   (a) Using a linear classifier and calculated weight for words evaluate subjectivity and emotional values of each sentence of a document separately.
   (b) On the basis of the preceding item calculate the features describing the entire document and add them to train feature matrix.
3. Train Random Forest algorithm [9] and calculate cross validation.

---

[1] http://nlp.stanford.edu/software/classifier.shtml
[2] http://text-processing.com/demo/sentiment/

To calculate the weights the author did not use the words with negation before them. For example, if there was a calculation for the weight of the word "good", then phrases like "not good" were ignored (though negations were taken into account when analyzing the emotionality of the sentences). In order to reduce the size of the vector space the Porter stemmer was used. In addition, the text went through another preprocessing procedure, namely the removal of stop words both from corpora used for calculation, laid to calculate weights, and a test collection (we removed all negation and intensifier words from the stop words list[3]).

Partitioning into sentences allows us to easily process different syntactic constructions. Implemented linear classifier considers the following lexical phrases described in works [1,7]: intensifiers, negations and oppositions. Oppositions in the English language are expressed using the word "but". The following approach was used: if two parts of the sentence opposed by using "but" have different values, then "but" was ignored. Otherwise, the weight of the opposed sentence part changed its sign.

In order to control the quality of the Random Forest [9] algorithm 10-fold cross validation were used: AUC, accuracy, F-measure. The Random Forest in Weka implementation ran with the following parameters: number of trees – 100, depth of trees – 10, the rest – by default. The author decided not to use the popular text analysis algorithm SVM, as the latter has the advantage for the problems with large number of features, when samples are almost linearly separable. Here we use only 11 features to describe the entire document sentiment (see next section), and Random Forest is better than SVM.

3   Feature Text Description

For a text description a set of features is used, each feature is calculated on a set of emotional and subjective values of the sentences constituting the text. Each of emotionality or subjectivity values is obtained by applying a linear classifier with weights Delta TFIDF, calculated on emotional and subjective corpora. There is another popular statistical measure — semantic orientation, which we do not use because it was experimentally proved that classifiers based on this measure perform worse than ones using Delta TFIDF (see [4,5,7]).

3.1   Delta TFIDF Measure

Before describing the method for analyzing the sentences we give a fundamental formula of the theory of information retrieval adapted for the sentiment analysis. Consider a balanced corpus, where positive documents number $|P|$ equals negative one $|N|$, then

$$Delta\,TFIDF = C_{t,d} \log_2 \frac{|N|}{N_t} - C_{t,d} \log_2 \frac{|P|}{P_t} = C_{t,d} \log_2 \frac{P_t}{N_t}. \tag{1}$$

---

[3] http://dev.mysql.com/doc/refman/5.5/en/fulltext-stopwords.html

Here, $C_{t,d}$ – number of times the word t appears in the document d, P_t and N_t – number of occurrences of term t in the positive and negative documents. When the numerator or the denominator equals zero, the logarithm value is undefined. To avoid this case Laplacian smoothing was used: we assume, that word occurred 1 more time than it actually did. Let denote as Delta IDF the fraction under the logarithm in (1). Delta IDF does not depend on the document. Consequently, it can be calculated in advance once for all the words in the corpus.

3.2   Feature Set

To assess emotionality or subjectivity we use a linear classifier $f(x) = \text{sign}(x, \mathbf{w})$, where the vector x is the characteristic vector of a sentence text, with i-th coordinate equals the number of occurrences of the i-th word in the text. Coordinates of the vector $\mathbf{w}$ represent the weight of the words in a text. The weight is calculated based on Delta IDF. To calculate the subjective/objective weight we use the corpus in which all the sentences were divided into two classes: subjective and objective. To calculate emotional weight we use tweets and the reviews from amazon.com.

As stated in Sect. 3, on the basis of emotional and subjective values of the sentences, following groups of attributes describing the whole text in general were calculated:

1. Number of positively, negatively and neutrally classified sentences.
2. Result of the linear classifier for the entire text.
3. Average number of words in positive and negative sentences.

Also, we duplicated these features for subjective sentences (except neutrally classified sentence number), because account subjectivity improves quality of sentiment analysis [4,6]. Result of linear classifier was added because addition new strong features should improve classification quality.

4   Numerical Experiments

To test described in the second section algorithm we used test collection[4] by B. Pang and L. Lee, consisting of 1000 positive and 1000 negative very detailed English reviews about films.

4.1   Description of the Sets for Calculation Delta IDF

We use 3 corpora for calculation Delta IDF dictionaries. One of these corpora was the database of reviews from amazon.com[5] on various topics (a total of 20 subjects and 250 000 reviews). Every review was rated on a five-point scale. Neutral reviews (mark 3) were not presented. It was natural to assume that

---

[4] http://www.cs.cornell.edu/people/pabo/movie-review-data/
[5] http://www.cs.jhu.edu/~mdredze/datasets/sentiment/unprocessed.tar.gz

Table 1. Results

	AUC	Accuracy(%)	F-measure(%)
StandartDictionary	0.92 ± 0.006	86.75 ± 0.39	86.67 ± 1.04
ShortDictionary	0.901 ± 0.006	83.55 ± 1.10	82.09 ± 1.01
LongDictionary	0.893 ± 0.008	81.55 ± 0.70	81.38 ± 1.21
TweetDictionary	0.756 ± 0.003	68.85 ± 3.00	68.69 ± 4.00

review with the mark greater or equal to 4 was positively evaluated, and with the point less than 3 was negatively evaluated. To test the hypothesis that short texts are the best to use for calculation Delta IDF dictionaries, this collection was divided into two parts: reviews with the length of more than three sentences and shorter reviews. Two obtained dictionaries we denote LongDictionary and ShortDictionary respectively (Table 1).

The second corpus[6] represents the database of tweets on various topics. Altogether there were more than 1,500,000 tweets, each of them was assessed as either positive or negative. Usually, tweets go through different preprocessing steps: correcting typos, bringing the intentional errors (like ≪coool≫) to a special form and so on. The test set contains almost no emoticons, acronyms and intentional errors, hence we decided not to make any preprocessing of tweets. We will mark the obtained dictionary as TweetDictionary.

The subjects of all previous corpora actually have no relation to the one of the test collection. It is necessary to check the algorithm, with Delta IDF, calculated on a corpus with the subject close to the test set. This would contribute to either testing the hypothesis that for the Delta IDF calculation corpora of any subject can be taken, or checking the quality of the suggested model by comparing with [4, 6]. Therefore, we took random 300 positive and 300 negative reviews from a test collection and constructed the third corpus. Denote dictionary Delta IDF, obtained in this corpus, as StandartDictionary.

4.2 Results of the Experiments

Using cross-validation of the model on the test collection of films and described above four corpora for Delta IDF calculation, the following results were obtained:

First of all, it should be noted that there is a considerable decline in the accuracy when TweetDictionary for Delta IDF calculation is used. It could be explained by the fact: expression of emotions in Twitter comes out with emoticons, intentional errors, acronyms, slang much more than the classical vocabulary.

Experimental results show that the suggested text model in the article almost is not inferior in accuracy to [4, 6]. Moreover, if we use the entire test collection for calculating Delta IDF, then the value of accuracy reaches 96, which is higher than [4, 6] and almost the same as [5]. The author believes that the approach in [5]:

---

[6] http://thinknook.com/twitter-sentiment-analysis-training-corpus-dataset-2012-09-22/

using the same sample either for cross validation or Delta IDF calculation —
is not quite correct. The right decision is recalculation of Delta IDF weights at
each iteration of cross-validation or using not intersected corpus for designing
dictionary with test collection corpus for Delta IDF calculation.

5   Conclusion

The developed model deals with texts like blogs, reviews from amazon, facebook
etc. It has two-level structure. At the first stage we calculate subjectivity and
emotional values of sentences comprising the text using linear classifier. On the
next stage we classify the text based on obtained statistical features. Model correctly handles negations, intensifiers, oppositions. It demonstrates high accuracy
on popular test collections.

The experiments showed that the model successfully works for cross-domain
sentiment analysis. The hypothesis "in order to calculate Delta TFIDF it is
better to use short texts" has been tested. Tweet-based dictionary caused considerable decreasing of accuracy. However, short review dictionary from amazon.com
showed better results than the same dictionary based on long reviews.

References

1. Liu, B.: Sentiment Analysis and Subjectivity - Handbook of Natural Language
   Processing, 2nd edn. Taylor and Francis Group, Boca (2010)
2. Pang, B., Lee, L., Vaithyanathan, S.: Thumbs up? Sentiment classification using
   machine learning techniques. In: Proceedings of EMNLP, pp. 79–86 (2002)
3. Pang, B., Lee, L.: A sentimental education: Sentiment analysis using subjectivity
   summarization based on minimum cuts. In: Proceedings of the ACL (2004)
4. Martineau, J., Finin, T.: Delta TFIDF: An improved feature space for sentiment
   analysis. In: Third AAAI Internatonal Conference on Weblogs and Social Media,
   May 2009, San Jose, CA (Preprint) (2009)
5. Paltoglou, G., Thelwall, M.: A study of information retrieval weighting schemes for
   sentiment analysis. In: Proceedings of the 48th Annual Meeting of the Association
   for Computational Linguistics, pp. 1386–1395. Uppsala, Sweden (2010)
6. Pang, B., Lillian, L.: A sentimental education: sentiment analysis using subjectivity
   summarization based on minimum cuts. In: Proceeding ACL '04 Proceedings of the
   42nd Annual Meeting on Association for Computational Linguistics Article No. 271,
   USA (2004)
7. Taboada, M., Brooke, J., Tofiloski, M., Voll, K., Stede, M.: Lexicon-based methods
   for sentiment analysis. J. Comput. Linguist. 37(2), 267–307 (2011)
8. Peter, D.T.: Thumbs up or thumbs down? Semantic orientation applied to unsupervised classification of reviews. In: Proceedings of the 40th Annual Meeting of the
   Association for Computational Linguistics (ACL), Philadelphia, pp. 417–424, July
   2002
9. Breiman, L.: Random forests. Mach. Learn. 45(1), 5–32 (2001)

Image Restoration Algorithm Based on Regularization and Adaptation

Tatiana Serezhnikova[1,2]([✉])

[1] Krasovsky Institute of Mathematics and Mechanics UB RAS, Ekaterinburg, Russia
[2] Ural Federal University, Ekaterinburg, Russia
sti@imm.uran.ru

Abstract. We propose to add a special summand (stabilizer) to the original Tikhonov regularization algorithm; this regularizer includes a specially adapted function to the solution characteristics for the problem of image restoration. This approach to approximation of non-smooth functions based on our new technique for choosing interpolation points. As a result, the approximate solutions have better accuracy and images become more deblured. Moreover, it becomes possible to keep small objects and contours in complex scenes by incorporation of background knowledge about their location or structure into the regularization procedure.

Keywords: Image reconstruction · Non-smooth solution · Adaptive method · Ill-posed problem · Tikhonov regularization · Fredholm integral equation · Numerical method

1 Introduction

Many important applications in medicine, science, space exploration, precision engineering, and so on have very stringent quality requirements on image restoration algorithms. In the paper, we consider images as two-dimensional functions $u(x, y)$ depending on x and y being coordinates on the plane. The value of u is called the intensity of the image at the point (x, y). If the values of x, y and u are taken from some finite set of discrete values, then the image is called digital. Digital image processing has a wide and varied field of applications. Algorithms for spatial image processing usually are more computationally efficient. Spatial image processing is described by the equation $A(u) = f$, where A is an operator over u in the neighborhood of points, $u = u(x, y)$ is the current image, $f = f(x, y)$ is the processed image.

The term spatial filtering is often used in the digital image processing. The main problem of this type of filtering is its sensitivity to noise. Filtering by Tikhonov regularization, together with a choice of suitable input parameters, proved its effectiveness, see [1–6].

An intrinsic feature of Tikhonov regularization is smoothing effect, so that the graph of the approximate function $u(x, y)$ has smoothed breaks and fractures.

© Springer International Publishing Switzerland 2014
D.I. Ignatov et al. (Eds.): AIST 2014, CCIS 436, pp. 213–221, 2014.
DOI: 10.1007/978-3-319-12580-0_22

If filtering methods smooth out breaks and fractures of function graphs, it may cause bad quality of processed images. Thus an important problem here is to construct a Tikhonov regularization method in order to improve the approximation of non-smooth (or even discontinuous) solutions of the operator equation $A(u) = f$.

The paper presents our method based on Tikhonov regularization. In our work we introduce an additional term in the basic Tikhonov regularization. This additional term is adapted to the characteristics of the reconstructed image, i.e. it exploits background knowledge. Thus our work provides the new contribution to the methods of numerical functional minimization.

Looking at the figures in the paper, one can see that our technique allows to compute the image reconstructions with the adaptation to background knowledge and to obtain debrlured images.

The paper is organized as follows. In Sect. 2 we describe regularization functionals and present the main idea of our technique. In Sect. 3 we describe the results of our experiments and demonstrate graphs for several model problems. In Sect. 4 we pay special attention to the more general form of the proposed regularizer and related experiments. The last section concludes the paper.

2   Mathematical Model for Image Deblurring

We consider the following two-dimensional Fredholm integral equation of the first kind:

$$Au \equiv \int\limits_0^1 \int\limits_0^1 K(x - \xi, y - \eta)u(x, y)dxdy = f(\xi, \eta). \tag{1}$$

In image reconstruction, the estimation of u from observation of f is referred to as the two-dimensional image deblurring problem ([5], p.64). In optics, u is called the light source, or object. The kernel function K is known as the point spread function (PSF), and f is called the blurred image. The image is often recorded with a device known as a charge-coupled device (CCD) camera.

We are interested in reconstruction methods for complex scenes containing various important small-size objects and corresponding to non-smooth solutions of Eq. (1). Using total variation, one can effectively reconstruct functions with jump discontinuities.

We construct the original method to solve problem (1). Let $A : U \to F$ be a linear operator, where U and F be linear normed spaces. Assume that the inverse operator A^{-1} is discontinuous, then the equation $Au = f$ is said to be ill-posed problem.

Abstract methods with full investigation of the convergence of regularization algorithms for this problem presented in [3,4].

The foundation of the regularization method is given by

$$\min\left\{\|A_h u - f_\delta\|_{L_2}^2 + \alpha\left(\|u\|_{L_2}^2 + J(u)\right) : u \in U\right\}, \tag{2}$$

where

$$J(u) = \int_D |\nabla u| \, dx, \tag{3}$$

where ∇u denotes gradient of the smooth unknown function u, $(u \in W_1^1(D))$, $J(u)$ is the total variation of the function u on D.

The practical implementation of this method requires minimization of a discretized version of the functional in (2). A completely discrete model may be obtained by truncating the region of the integration in (1) to be the union of the small squares $h \times h$, $h = 1./n$ and then by applying the midpoint quadrature to (1). So, Eqs. (2) and (3) are reduced to

$$\min\left\{ \sum_{k,l} h^2 \left[\sum_{i,j} h^2 K(y_k - t_i, y_l - s_j) u(t_i, s_j) - f_{k,l} \right]^2 \right.$$
$$\left. + \alpha \sum_{i,j} h^2 \left\{ u_{i,j}^2 + \left[\left(\frac{u_{i,j} - u_{i,j-1}}{h} \right)^2 + \left(\frac{u_{i,j} - u_{i-1,j}}{h} \right)^2 \right]^{\frac{1}{2}} \right\} \right\}, \tag{4}$$

Composition of two well-known approaches, the Tikhonov variational representations (1)–(4) and the version of the iterative technique, containing additional parameters $\beta_{i,j}$, $0 < \beta_{i,j} < 1$, see [2,3]:

$$\mathbf{u}^k = \arg\min \left\{ \Phi_N^\alpha(\mathbf{u}) + \sum_{i,j} \beta_{i,j} (u_{i,j} - u_{i,j}^{k-1})^2 : \mathbf{u} \in R^N \right\} \tag{5}$$

is the main contribution of our paper.

Here, $N = n^2$, $\Phi_N^\alpha(\mathbf{u})$ is functional in (5).

We use the iterative subgradient method in order to compute $\mathbf{u}^k$ defined in (5)

$$\mathbf{u}^{k,\nu+1} = \mathbf{u}^{k,\nu} - \lambda_k \frac{\mathbf{v}^{k,\nu}}{||\mathbf{v}^{k,\nu}||}, \quad \nu = 0, 1, 2, \dots, n_k, \tag{6}$$

where $\Phi_N^{\alpha,\beta}(\mathbf{u}^{k,\nu})$ is the functional in (5), $\mathbf{v}^{k,\nu} \in \partial \Phi_N^{\alpha,\beta}(\mathbf{u}^{k,\nu})$ and $\partial \Phi$ is an arbitrary subgradient of the functional Φ; $\{\lambda_k\}$ are parameters for the iterative processes control actions, see details in [3].

3 Numerical Experiments

3.1 Motivation

In the beginning of the work, we used in (5) the constant $\beta, \beta_{i,j} = \beta$, in every mesh points of discrete models. One can see in [3] that our technique showed a good ability to preserve local extreme points and jumps of functions.

Namely, the failure of calculations for the first two-dimensional model motivated us to change the way of $\{\beta_{i,j}\}$ selection. In [2] we set

$$\begin{cases} \beta_{i,j} = 10^{-10} & \text{for the points, where } u_{\text{true}} = \max\{u_{\text{true}}\}; \\ \beta_{i,j} = 0, & \text{where } u_{\text{true}} < \max\{u_{\text{true}}\}. \end{cases} \tag{7}$$

3.2   Comparisons with Total Variation Regularization Solutions for Image Deblurring Reconstructions

We have compared numerical results obtained for setting a zero parameter β, $\beta \equiv 0$, and numerical results for adapted $\{\beta_{i,j}\}$ in (5). One can see that our numerical Algorithm (2)–(6) with the parameter $\beta \equiv 0$ is a variant of the total variation regularization technique in two-dimensional case. In order to demonstrate the qualitative difference between numerical results obtained for setting $\beta \equiv 0$ and for setting adapted $\{\beta_{i,j}\}$, we present plots of two blurred image reconstructions in this subsection. It is very interesting to compare performance of these two solution techniques. An overall cost comparison is difficult to carry out, since it depends on many factors. Our numerical results let us to produce the visual comparisons of these two solution techniques for two tests, see Figs. 1, 2, 3, and 4.

In order to compare the results, we present the plot of the true image diagonal and plots of the corresponding reconstruction diagonals in Fig. 2.

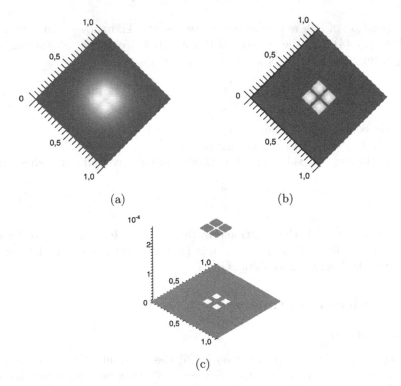

(a)   (b)

(c)

Fig. 1. The image reconstruction for the first model. (a) Plot shows blurred image. (b) Plot shows reconstruction from data (a) for adapted $\{\beta_{i,j}\}$ (see (8)). (c) Plot shows the true object in the three-dimensional coordinate system.

For the model image in Fig. 1(c), the diagonal $u_{\text{true}}(x, x)$ of the true image is given by

$$u_{\text{true}}(x, x) = \begin{cases} 2.5 \times 10^{-4}, \ 0.391 \leq x < 0.484, \\ \qquad\qquad 0.531 \leq x < 0.625; \\ 0. \ , \qquad \text{otherwise.} \end{cases} \tag{8}$$

In Fig. 2, we have visual comparisons of two solution reconstructions for blurred image in Fig. 1(a). One can see that the algorithm with adapted $\{\beta_{i,j}\}$ preserves jumps of the function much better (compare Fig. 2(b) and Fig. 2(a)).

The base configuration for the true model in Fig. 3(a) is taken from [5]. In Fig. 3(a) the most interesting detail is the white triangle. Figure 4 demonstrates visual comparisons of two reconstructions for this triangle.

One can see in Fig. 4(b), that due to adapted $\{\beta_{i,j}\}$, the plane triangle reconstruction is better (compare with Fig. 4(a)).

4 $\beta(x, y)$ Construction Using Approximate Solutions

We propose addition of the special summand into classic regularizing functionals, containing regularization parameter α too, see (5).

In this paper, we propose a more general form of the special summand, which we put in classical regularizing functionals. For the regularizer I^β, which we describe, we put $I^\beta = \int_Q \beta(x, y)[u(x, y) - u^k(x, y)]^2 \, dxdy$, where

$$\begin{cases} \beta(x, y) = \beta_0 \times u_{\text{max}}, & \text{for} \ \ (x, y) \in Q, \\ \beta(x, y) = 0, & \text{for} \ \ (x, y) \in \Pi/Q, \quad \text{where} \end{cases} \tag{9}$$

$u_{\text{max}} = \max\{u_{\text{true}}(x, y), (x, y) \in \Pi = [0, 1] \times [0, 1]\}, \quad (x, y) \in Q \Leftrightarrow u(x, y) = u_{\text{max}}.$

For the model image in Fig. 6, $\beta_0 \approx 10^{-5}$ and the $u_{\text{true}}(x, y)$ is given by

$$u_{\text{true}}(x, y) = \begin{cases} 2.5 \times 10^{-4}, \ 0.391 \leq x, y \leq 0.610, \\ \qquad 0. \ , \qquad \text{otherwise.} \end{cases} \tag{10}$$

In Fig. 5, plots show the third image model. In Fig. 5(a), the small square points set is the point set Q in (9). If we set Γ_Q to be the Q bound, then Γ_Q is the small square bound in (9).

In Fig. 6 plots show the reconstruction of the image in Fig. 5 (a).

Now, using data in Fig. 7, we describe the idea how it is possible to use an approximation solution $\tilde{u}(x, y)$ instead of $u_{\text{true}}(x, y)$ in (9).

The image in Fig. 7 (a) shows the small square reconstruction. This reconstruction has been calculated for the parameter function $\beta \equiv 0$.

Plots in Fig. 7 (b) show projections of two graphs: the first graph (the parallel line to the coordinate axis) is the projection of the true small square image; the second graph is the graph of the approximation of image for $\beta \equiv 0$. Let $\tilde{u}^0(x, y)$ be the approximation of image reconstruction for $\beta \equiv 0$. Then, we propose to use $\tilde{u}^0(x, y)$ instead of $u_{\text{true}}(x, y)$ in (9).

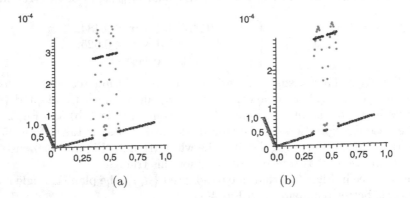

Fig. 2. Plot of the true image diagonal (8) and their reconstructions. (a) Case of $\beta \equiv 0$. (b) Case of $\{\beta_{i,j}\}$ given by (7).

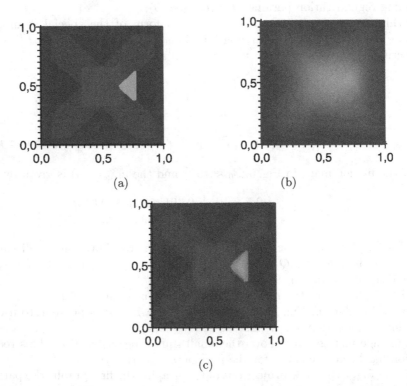

Fig. 3. Plots show the 128×128 simulated satellite in Fig. 3(a), the blurred image in Fig. 3(b) and the reconstruction image in Fig. 3(c) for adapted $\{\beta_{i,j}\}$.

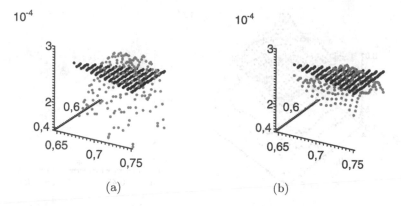

(a)   (b)

Fig. 4. Plots demonstrate visual comparisons of two reconstructions for the image model in Fig. 3(a). (a) Plots show the true solution triangle part and the reconstruction of the triangle part for $\beta \equiv 0$; (b) Plots show the true solution triangle part and the reconstruction of the triangle part for the adapted $\{\beta_{i,j}\}$.

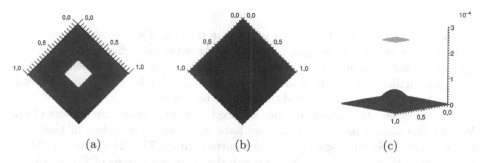

(a)   (b)   (c)

Fig. 5. (a) Plot demonstrates the true image; (b) Plot demonstrates the observed image; (c) Plot demonstrates the image (a) and the image (b) as two-dimensional functions.

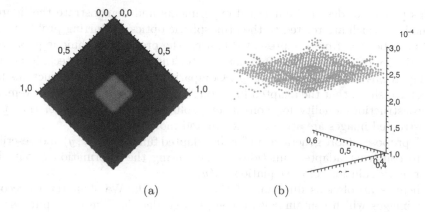

(a)   (b)

Fig. 6. (a) Plot shows the reconstructed image for the adapted $\beta(x, y)$; (b) Plots show the main true detail (the inside square) and it is reconstruction for the adapted $\beta(x, y)$.

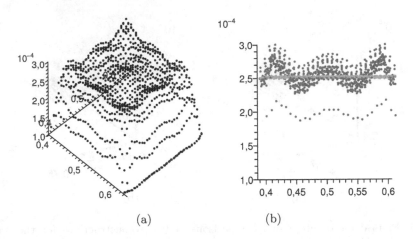

(a)   (b)

Fig. 7. (a) Plot in Fig. 7 (a) shows the graph of the small square reconstruction for $\beta \equiv 0$; (b) There are projections of two plots: the true small square image and the small square approximate image for $\beta \equiv 0$.

In Fig. 7(b) points of the $\tilde{u}^0(x, y)$ are located at the $4 \times h_1$-width band, were h_1 is the axis unit. And the $u_{\max}$ is located at the same band. So, we can take the approximation $\tilde{u}_{\max}$ such that $\Delta = |\tilde{u}_{\max} - u_{\max}| < 2 \times h_1$, for which $\Delta \approx 8\%$ of $u_{\max}$. Further, we use the numbers $\{\tilde{u}_{\max}, \tilde{u}_{\max} \pm 2 \times h_1, \tilde{u}_{\max} \pm h_1\}$ instead of $u_{\max}(x, y)$ and therefore, we obtain the better results.

In case of the Γ_Q approximation, we need to approximate the corners of Q. We can use similar arguments that we have mentioned above but in this case $h_2 \approx 1./128$). We can approximate each corner within 7%. Then, we should to use no more than 3 or 4 good lines for suitable approximations of Γ_Q in (9).

5   Conclusions

In this paper, we describe numerical experiments and demonstrate the figures for models, which are related to the atmospheric optics deblurring problem.

As can be seen from the presented figures, the proposed technique preserves discontinuity of reconstructed functions better than the original technique due to adaptations of the function $\beta(x, y)$. Comparison with the reconstructions for $\beta(x, y) \equiv 0$ show, that the adapted function $\beta(x, y)$ ensures crucial improvement of reconstructions quality for non-smooth solutions. Finally we see, that the reconstructed images are accurate enough and more deblurred.

We propose a more general form for the adapted functions $\beta(x, y)$ and describe how to construct adapted functions $\beta(x, y)$, using the information about the approximate solutions of the equations $A(u) = f$.

There is an obvious direction of the future work. We shall try to reconstruct images which contain several complicated details. The main purpose of the research remains to improve reconstructions of function jumps and to obtain more deblured images using background knowledge.

Acknowledgments. The author would like to express special gratitude to Prof. V.V. Vasin from the Institute of Mathematics and Mechanics UB RAS. This work was supported by Russian Foundation for Basic Research, project no. 12-01-00106. I would like to thank the colleagues from AIST Program and Organizing Committees for their helpful advice and guidance in the paper preparations and supported by the Program of Presidium RAS N 15 (project 12-P-1-1023).

References

1. Gonzalez, R.C.: Digital Image Processing, 3rd edn. Prentice Hall, New Jersey (2008)
2. Serezhnikova, T.I.: An algorithm based on the special method of the regularization and the adaptation for improving the quality of image restorations. Univer. J. Comp. Math. **2**(1), 11–16 (2014)
3. Vasin, V.V., Serezhnikova, T.I.: A regularizing algorithm for approximation of a nonsmooth solution of Fredholm integral equations of the first kind. J. Vich. Tech. **15**(2), 15–23 (2010)
4. Vasin, V.V., Serezhnikova, T.I.: Two steps method for approximation of a nonsmooth solutions and nosy image reconstructions. J. Avt. Tel., 2, 126–135 (2004)
5. Vogel, C.R.: Computational methods for inverse problems. SIAM, Rhiladelphia (2002)
6. Chen, X., Ng, M.K., Zhang, C.: Non-Lipshitz l_p-regularization and box constrained model for image reconstruction. IEEE Trans. Image Process. **21**(12), 4709–4721 (2012)

Cognitive Semiotic Model for Query Expansion in Question Answering

Alexander Sirenko[1,4]($\boxtimes$), Galina Cherkasova[2,4], Yuriy Philippovich[3,4],
and Yuriy Karaulov[3,4]

[1] Moscow State University of Printing Arts, Pryanishnikova str., 2A, Moscow, Russia
[2] The Institute of Linguistics, Russian Academy of Sciences,
1 bld. 1 Bolshoi Kislovsky lane, Moscow, Russia
[3] Bauman Moscow State Technical University, 2nd Baumanskaya str., bld. 5,
Moscow, Russia
[4] The V.V. Vinogradov Russian Language Institute of the Russian Academy
of Sciences, ulitsa Volkhonka, 18/2, Moscow, Russia
{sirenko,cherkasova,y_philippovich,karaoulov}@it-claim.ru

Abstract. Query expansion improves performance of informational
retrieval stage in question answering pipeline. We state the benefits
of a personalized and autonomous query preprocessing and automate
a semiotic model to achieve such properties. The model operates as a
context-sensitive weighted grammar, along with the algorithm to apply
production rules allowing approximate matching. The semiotic model is
packed into a regression model to predict relevant terms for a query.
ROC-analysis evaluates the regression model and helps to choose the
optimal cutoff level. We compare ranking of terms by regression model
and ranking based on an external informational retrieval system.

Keywords: Semiotic modeling · Cognitive experiments · Grammar ·
Query expansion · Regression · Question answering

1 Introduction

Question Answering (QA) is a well established task, supported by the specific
tracks of Text REtrieval Conference (TREC). Recently published solutions follow the steps of QA pipeline approach [1]: (a) preprocess a query; (b) acquire
relevant documents/passages to the query (information retrieval, IR); (c) extract
information to get possible answers from relevant documents; (d) rank extracted
answers to choose the best of them.

The QA pipeline finds answers in a collection of unstructured texts, therefore
the IR bounds the performance of the whole QA. Different techniques change a
query to increase the recall of the IR. In this research we focus on the expansion
of the query with the relevant terms within the QA pipeline.

Since the Internet is the most popular way to deliver informational services,
we take into account two trends: personalization of services and widely used

© Springer International Publishing Switzerland 2014
D.I. Ignatov et al. (Eds.): AIST 2014, CCIS 436, pp. 222–228, 2014.
DOI: 10.1007/978-3-319-12580-0_23

powerful personal devices to access the web. Currently users share private data (history of queries, location, messages) with a web-search system to get more relevant search results. If a query preprocessing operates locally and doesn't communicate with an external system, the user may allow to access personal information. Such local preprocessor should be build without expensive structured data.

We suggest to build a cognitive semiotic model upon an associative network and a set of cognitive entities [2]. Psycholinguists of the "Russian Personality" school described the cognitive model and interviewed people to collect empirical data. It leads us to the cognitive model of an "average" user.

2 Related Work

State of the art QA systems use different approaches to expand a query: mutual information technique [1,3,4]; statistical machine translation [5]; relevance feedback technique, which extends a query based on the first results from IR [6]. These techniques need a QA corpus or access to an external IR-system to expand a query. It contradicts the ideas of personalization and autonomy, therefore we propose to apply the cognitive model to expand a query. Cognitive models are usually task-specific (transformational grammar of Noam Chomsky [7]) or the opposite - general frameworks (ACT-R [8], Soar [9]). The chosen cognitive model accepts a sentence in natural language and outputs a set of terms. The model agrees with a query expansion task, addresses personalization and autonomy.

3 Data

The associative experiment was conducted using the method of free associations, where the respondent produces the first association for a given word. We construct a network from the lemmatized associative dictionary of Russian [10]. The network contains 63,700 nodes, 394,000 connections in surface form and 29,290 nodes, 132,456 connections in lemmatized form. An example of raw associations: (road, far, 28), (road, long, 19), (road, to home, 15), where the last number - strength of association. The cognitive experiment methodically copies the game "crossword", where a respondent guesses a term by description. The question-answer pairs with additional attributes form cognitive entities (cognems). A cognem has 5 components: *Sign, Meaning, Form_of_meaning, Function, Domain*. *Sign* stands for a word to be guessed. *Form_of_meaning* defines the structure of a statement in *Meaning* (definition, metaphor etc.). *Function* shows necessity of cognem for day-to-day life. *Domain* locates the cognitive entity relative to other cognems of the respondent's knowledge (literature, army, instruments etc.) For example: ("accent", "A way of pronunciation by a speaker in native language", "definition", "necessary", "language"); ("tree", "It has green leaves", "description", "necessary", "forest").

The original set of 18,300 cognems covers 55 variants of *Form_of_meaning* and 231 variants of *Domain*. The raw set of cognems contains 11,000 unique signs and 31900 terms in *Meaning*. Lemmatized cognems include 3,974 terms in *Sign* and 22,208 terms in *Meaning*.

4   Building a Query Expansion Preprocessor

The query preprocessor here is a classifier of relevant terms to a query with the cognitive model inside to provide dependent variables. To construct the preprocessor, we:

1. construct a context-free grammar G from the associative network and a dictionary of synonyms [11];
2. construct a weighted context-sensitive grammar G_{Ext} from G and cognems;
3. generate possible expansion terms by G_{Ext}, train regression model, choose optimal cutoff and evaluate classifier.

4.1   Context-Free Grammar

The elements of the associative network: S - stimulus, R - reaction, SR - stimulus and reaction, $prob_{ij}$ - probability of encountering the association from node i to j in a free associative experiment. The associative network and synonyms are incorporated into context-free grammar: nodes of associative network and synonyms form grammar's symbols. Derivation in G is performed as in Markov chain: we define probability to get next sentence based on probability of current sentence and probability of a rule. Associations form rules from a stimulus node to a reaction node with probability $prob_{ij}$. Synonymous pairs form rules with $prob_{ij} = 1$.

G applies rules with the biggest probability to generate sentences from the query. An applied rule $(t_k, (t_m...t_n), prob_{k,m..n})$ replaces term t_k in the query with terms $(t_m...t_n)$, generating new sentence S_i. Probability to generate S_i is a multiplication of applied rules' probabilities. For example: Query $S_0 = (t_1, t_2, t_3)$ after applying rules $r_1 = (t_1, t_4, prob_{1,4})$ and $r_2 = (t_4, t_5, prob_{4,5})$ will generate: $S_1 = (t_4, t_2, t_3)$ with probability $prob_{1,4}$; $S_2 = (t_5, t_2, t_3)$ with probability $prob_{1,4} * prob_{4,5}$. If same S_i could be generated by different sequences of applied rules, it's probability aggregates multiplied probabilities for each sequence.

4.2   Context-Sensitive Grammar

For each $cognem = (sign, meaning, domains)$ we make a grammar rule $P_{Ext} = (meaning, sign, \emptyset)$. *meaning* and *sign* parts of a P_{Ext} are lemmatized terms of *meaning* and *sign* parts of a *cognem* with the original order. Rules $\{P_{Ext}\}$ form conjointly with the G a context-sensitive grammar G_{Ext}. We incorporated Damerau-Levenshtein distance with suffix trees into an automaton in order to match approximately the *meaning* part of P_{Ext} with the query [12]. The automaton defines inverted cost $rInvCost$ of cognem-based rules $\{P_{Ext}\}$ applied to some input sentence. $rInvCost$ of context-free rules in G_{Ext} equal to $prob_{ij}$ of G.

$rInvCost$ replaces $prob_{i,j}$ in calculation of generated sentence's probability. Since we mixed probability from G with euristic value $rInvCost$ for $\{P_{Ext}\}$, sentences generated by G_{Ext} have not probability but weight - inverted cost of a sentence $sInvCost$.

Sentences generated by G_{Ext} contain terms, that are not present in the original query. These terms form a set of possible expansion terms $expTerms$. Each new term has properties $(t_i, tInvCost_i, tUsage_i)$, where $tInvCost_i$ - aggregated $sInvCost$ of all generated sentences with t_i; $tUsage_i$ - ratio of query terms, used in derivation of t_i to count of terms in the query.

If G_{Ext} runs on "Capital [of the] largest oil-producing country" it transforms the query with the cheapest associative rule (capital $\rightarrow$ Soviet) to "Soviet largest oil-producing country". Triplet ("Soviet", $tInvCost^{-1}_{Soviet \rightarrow capital}$, 0.25) joins $expTerms$. The transformed sentence joins the set for further derivation. Afterwards G_{Ext} applies the cheapest context-sensitive rule to the original query and marks the original query as processed. At this point we have 1 or more terms in $expTerms$ and 1 or more sentences to derive from. G_{Ext} chooses not processed sentence with the biggest $invCost$ and runs on it. G_{Ext} stops when $expTerms$ reached predefined size. See results in Table 2.

4.3   Classifier of Relevant Terms

The linear regression model predicts relevance of term t_i to $query$ as follows: $Y(t_i) = b_0 + b_1 * tInvCost_i + b_2 * tUsage_i$. We train and test it with true positive pairs $(meaning, sign)$, taken from cognems.

Considering the abbreviations TP (true positive), FN (false negative), TN (true negative), FP (false positive), the regression model has properties: (a) Sensitivity $Se = \frac{TP}{TP+FN} * 100\%$; (b) Specificity $Sp = \frac{TN}{TN+FP} * 100\%$; AUC - area under the ROC-graph's curve. ROC-analysis uses a graph $Se = F(1 - Sp)$. AUC positively relates to the quality of the regression model. With $cutoff = argmax_k(Se_k + Sp_k)$ the regression model classifies derived symbols on relevance to the query. We could use $Domain$ property of the cognitive entities to improve the quality of classification [13].

5   Results

Our final grammar G_{Ext} contained: 28,000 symbols, 123,000 rules and 3,900 synsets. Symbols inside a synset replace each other with inverted cost 1. Running the model against the query "[The] Capital of the largest oil-producing country" produces a list of symbols which contains the correct answer "Moscow" in position 2 and less relevant symbols "city", "Russia" in positions (7, 1) (see Table 2).

State of the art QA systems are built and tested on English corpora. To evaluate the impact of the query preprocessor we should compare the performance of the QA with and without preprocessing, or to check if additional terms improve the retrieval of documents with a correct answer. Since the cognitive model

Table 1. AUC evaluated on different test queries

Form_of_meaning	Terms in a query	Count	AUC of queries tested
Description	–	199	0.86
Definition	–	196	0.90
–	[5–10]	192	0.8
–	[1–5]	200	0.87

Table 2. Ranking for query "Capital of the largest oil-producing country"

Symbol found by model	tInvCost (rank)	tUsage (rank)	DocsCount ($< query > + < symbol >$)	Frequency (symbol)	DocsCount /Frequency
city (gorod)	2.271 (7)	5/6 (2)	3600	573.4	6.28
Moscow (Moskva)	3.153 (2)	5/6 (2)	3600	633.5	5.68
road (doroga)	2.604 (4)	5/6 (2)	1390	330.1	4.21
big (bol'shoi)	2.374 (5)	5/6 (2)	3640	944.4	3.85
Russia (Rossija)	3.687 (1)	4/6 (3)	3580	952.0	3.76
factory (zavod)	2.348 (6)	5/6 (2)	610	164.0	3.71
alley (alleja)	2.025 (8)	6/6 (1)	56	16.0	3.50
red (krasnij)	2.786 (3)	5/6 (2)	620	240.5	2.58

was not built with English data, the direct comparison has not been done yet. We use two methods to get insights into its performance: consider model as a regression model of relevance expansion terms to a query; compare ranking of expansion terms by the regression model with statistics from IR on a query $< original\_query + expansion\_term >$.

In the first method we use two sets: 2,500 cognems to train the regression model (see section Method) and 1,342 cognems to test it. Here cognems are used as pairs $(query, answer)$. If we train or test the model with a query from some cognem, the corresponding rule from G_{Ext} is switched off. See Table 1 for the estimated AUC for different subsets of test queries. Values above 0.5 mean the regression model performs better than random, while values 0.8–0.9 are quite good characteristics.

In the second method we use the hypothesis from the Relevant Feedback technique: the relevant terms occur in documents with the query more often than non-relevant ones. The set of top ranked terms should persist for both rankings: based on IR co-occurrence and on the regression model. For IR the ranking criteria will be $rcoef_{s,q} = \frac{docsCount(query \cup symbol)}{Freq_{symbol}}$: $Symbol$ collocates with $Query$ inside documents AND $Symbol$ is rare in texts.

We compose a query $< original\ query > AND < symbol\ suggested\ by\ model >$ for the search engine Bing.com to acquire $docsCount(query \cup symbol)$ (See Table 2). The results ranked by $DocsCount/frequency$ contain relevant answers in positions (1, 2, 5) which are closer to top spots. Note that the completely irrelevant answer "red" goes down in rank and "alley" is not in the bottom position due to low frequency.

6   Conclusion

Query expansion is often used to improve efficiency of a QA pipeline. We propose to use a cognitive semiotic model as a query preprocessor to achieve personalization and autonomy of expansion. The model is built upon the associative network and the set of cognitive entities. It operates as a context-sensitive grammar with approximate matching of a left part of a rule and a query. To suggest expansion terms for the query, the grammar is packed into a regression model of relevance of the terms to the query. We evaluate efficiency of the regression model through ROC-analysis and AUC. The model's top-ranked terms agree with the co-occurrence statistics from IR.

Acknowledgements. The research is supported by the grant №12-04-12039v of the Russian Humanitarian Scientific Fund and the grant №NSH-5740.2014.6 from the President of Russia.

References

1. Bilotti, M.: Query Expansion Techniques for Question Answering. Department of Electrical Engineering and Computer Science, Massachusetts Institute of Technology (2004)
2. Philippovich, Yu.N., Philippovich, A.Yu.: Dynamic infocognitive model of verbal consciousness. Neyrokomp'yutery: sozdanie i ispol'zovanie 1, 13–22 (2013) (in Russian)
3. Berger, A., Caruana, R., Cohn, D., Freitag, D., Mittal, V.: Bridging the lexical chasm: statistical approaches to answer-finding. In: Proceedings of the 23rd Annual International ACM SIGIR Conference (2000)
4. Komiya, K., Abe, Y., Morita, H., Kotani, Y.: Question answering system using Q & A site corpus query expansion and answer candidate evaluation. SpringerPlus 2, 396 (2013)
5. Riezler, S., Vasserman, E., Tsochantaridis, I., Mittal, V., Liu, Y.: Statistical machine translation for query expansion in answer retrieval. In: Proceedings of the 45th Annual Meeting of the Association for Computational Linguistics (ACL'07) (2007)
6. Derczynski, L., Wang, J., Gaizauskas, R., Greenwood, M.: A data driven approach to query expansion in question answering. In: Proceedings of Coling 2008, pp. 4–41. Coling 2008 Organizing Committee, Manchester (2008)
7. Chomsky, N.: The Minimalist Program. Current Studies in Linguistics Series, vol. 28, p. 420. MIT Press, Cambridge (1995)
8. Anderson, J.R., Bothell, D., Byrne, M.D.: An integrated theory of the mind. Psychol. Rev. 111, 1036–1060 (2004)
9. Lehnman, J.F., Laird, J., Rosenbloom, P.: A Gentle Introduction to Soar, an Architecture for Human Cognition. University of Southern California, Information Sciences Institute (1996)

10. Philippovich, A.Yu.: Automatized system for scientific research of associative experiments. Voprosy psikholingvistiki **6**, 143–153 (2007) (in Russian)
11. Abramov, N.: Dictionary of Russian Synonyms and Phrases with Similar Meaning, p. 7. Russkie slovari, Moskva (1999)
12. Sirenko, A.: Using of Non-determined Finite Automata for Approximate Substrings Matching in Context-Sensitive Grammar. Sbornik tezisov dokladov obshcheuniversitetskoy nauchno-tekhnicheskoy konferentsii "Studencheskaya nauchnaya vesna - 2011", p. 310. BMSTU n.a. Bauman, Moscow (2011) (in Russian)
13. Philippovich, Yu.N., Sirenko, A.V.: Classification of associative network's nodes: to build classifier. Izvestiya vysshikh uchebnykh zavedenii, Problemy poligrafii i izdatel'skogo dela, vol. 6 (2012) (in Russian)

Clause-Based Approach to Extracting Problem Phrases from User Reviews of Products

Vladimir Ivanov[1,2]([✉]) and Elena Tutubalina[1]

[1] Kazan (Volga Region) Federal University, Kazan, Russia
{nomemm,tutubalinaev}@gmail.com
[2] Institute of Informatics, Tatarstan Academy of Sciences, Kazan, Russia

Abstract. This paper describes approaches to problem-phrase extraction from user reviews of products. The first step in problem extraction is to separate sentences with problems from all others. We propose two methods to problem extraction from such sentences: (i) a straightforward algorithm that does not split sentence into clauses and (ii) an improved clause-based algorithm. We claim that both approaches improve the classification performance compared to machine-learning algorithms.

Keywords: Text classification · Information extraction

1 Introduction

Companies may correct some issues in their products, either services or devices, using customer feedback. Usually a defective product causes negative reviews. Such feedback from users allows a company to improve a product or a service only if the company tracks a particular device and the corresponding problems.

In this paper, we consider the task of extracting problem phrases from user reviews of products. In general, the task of problem-phrase extraction can be seen as a particular instance of the text categorization problem, in which only two classes are possible: the class with problems and the class without any problems. Example 1 contains a problem phrase. Example 2 does not contain problems.

1. I have been having problems with my printer for a few months as it has been spitting out pages that only had symbols.
2. I would recommend this printer to everyone who does not print in large volumes **but** wants good copies nevertheless.

The task determines whether each sentence contains a problem or not. However, it is difficult; one class is not completely the opposite to another. In Example 3 problem can be implied in the past; problems can occur because of the user (cf. Example 4). We consider such cases as mentions of *implicit*[1] problems.

3. I have only had one problem with this product, and that was years ago.

---

[1] The formal definition of the problem is difficult to determine.

© Springer International Publishing Switzerland 2014
D.I. Ignatov et al. (Eds.): AIST 2014, CCIS 436, pp. 229–236, 2014.
DOI: 10.1007/978-3-319-12580-0_24

4. I knew I needed to replace my black ink because of a message on my computer, **but** I did not remember which one and I could not buy a cartridge.

Some sentences from user reviews are *composite*. Conjunction words provide the necessary transition between two ideas in sentences 1–4. Conjunctions help to understand which of two ideas is more important. In this paper, we compare the performances of two approaches with machine-learning methods.

The rest of the paper is organized as follows: Sect. 2.1 presents a straightforward approach to extracting problem phrases; Sect. 2.2 describes an improved clause-based approach. In Sect. 3 we compare both approaches to each other on a manually constructed corpus of user reviews. Section 4 describes related work in the area of problem-phrase extraction or semantic analysis of contrast in natural language. Finally, Sect. 5 presents our conclusions and possible future extensions to this work.

2   Two Approaches for Problem-Phrase Extraction

The basic idea of both approaches we developed is to separate sentences with problems from all others. Obviously, this classification task might be solved by a number of supervised learning methods that make use of bag-of-words model. We have run such experiments and compared results to our original dictionary-based methods which we present in this section.

2.1   Straightforward Approach

The first algorithm evaluates simple conditions to decide whether a sentence contains a problem or not. The algorithm uses the following dictionaries: Action, NegativeWord, Negation, and ProblemWord. The Action dictionary contains all verbs that express some action, because a problem phrase may be presented as a negation of an action. The Action dictionary contains about 7,990 words. The NegativeWord dictionary contains a set of negative terms from a sentiment lexicon collected by Hu and Liu [1] and includes about 4,780 words. The Negation dictionary contains common negations (e.g. *not, n't, never* etc.). Finally, we include several types of words in the ProblemWord dictionary. The ProblemWord dictionary contains direct problem indicators (words such as *crash, break, reboot*), indirect problem indicators (words such as *tech support, sometimes, return*) and "not happening" problem words. The ProblemWord dictionary are manually created. A set of so-called not-happening problem words is also included in the Action dictionary. This subset contains specific words, such as *resolve, work, connect*. The negation of these words denotes a problem and may guarantee the problematical character of a sentence. Some words from the Action dictionary do not fall under this condition. Further, we describe steps of the algorithm.

Step 1. The algorithm looks for verb phrases headed by an action verb with a negation. If the sentence has an action word from the Action dictionary and a related word from the Negation dictionary, we extract the root phrase (S*) from the sentence and mark the sentence as a problem sentence.

Step 2. The algorithm looks for phrases in the sentence's syntactic tree[2] headed by a problem word. Then the algorithm looks if a selected phrase also contains a negation. If the sentence has a problem word from the ProblemWord dictionary without a related word from the Negation dictionary, we extract the root phrase (S*) from the sentence and mark the sentence as a problem sentence. We also mark the sentence by a problem label if the sentence contains a "not happening" problem word from the dictionary with a negation.

Step 3. The algorithm determines whether the sentence contains a problem using the results of step 1 and step 2.

Performance of this approach is discussed in Sect. 3.

2.2 Clause-Based Approach

We claim that most but-conjunctions change the meaning of a sentence. To illustrate this, we define three types of sentences and show that each sentence with a but-conjunction has two (opposite) semantic focuses.

Type-A. The first part of the sentence has a positive meaning, and the other part (after *but*) describes events that are different than expected.

Type-B. One part of the sentence confirms a problem description, and the other part (before or after *but*) denies an unpleasant situation or a problem.

Type-C. All parts of the sentence have similar information about situation. The but-conjunction is used to add something. All parts of the sentence have a problem descriptions or none of the parts have problems.

Type-A is illustrated in sentences 5 and 6. The first parts of these examples do not have any problem, but the other parts belong to a type of "denial of expectation" interpretation [2]. In example 6 the first part "great product" implies something that the second clause "cd does not work" contradicts. Type-A sentences have problems.

5. I have always been happy with hp products; and I love the new Photosmart Plus Printer, **but** I am extremely disappointed with the ink cartridges.
6. Great product, **but** CD does not work.

Type-B is given in Examples 7 and 8. The first parts of these sentences describe difficulties, but the second parts of the sentences express positive opinions. In 8 there is a contrast between the problem phrase "Printing is not fast" and the no-problem clause "that can be used", which relates to a different meaning in the sentence. Type-B sentences do not have problems with devices.

7. Admittedly I haven't had it long, **but** so far I love it.
8. Printing is not fast, **but** the variety of paper sizes that can be used for drawings and technical documents more than makes up for the speed.

---

[2] In the algorithm parsing step we use the Stanford Parser (http://nlp.stanford.edu/software/lex-parser.shtml).

Type-C is illustrated in 9. These sentences are more easily analyzed. In 9 we simply recognize that the "printer could not read" and "I could not clear the message"; these clauses only differ by subject of the phrase and by the verb, not by semantic meaning. Sentence 9 contains a problem.

9. The only thing I didn't like was when the printer could not read "print from an online source", it let me know, **but** I could not clear the message without turning it off and starting it again.

The interpretation of the but-conjunction contains semantic contrast in the problem-phrase extraction. Our improved approach consists of decomposing phrase extraction into a combination of simple decisions. In sentences with *but*, we detect the problem depending on the clause's meaning. There is a description of the improved clause-based algorithm.

Step 1. We define IP, DP, ¬DP, X, N values. The IP and DP represent two types of problem phrases. IP indicates an indirect problem phrase headed by one of action verbs with negation. Similarly, DP represents a direct problem phrase headed by one of problem words, while ¬DP indicates a phrase containing a problem word with a related negation (for this step we use a parsing method similar to a straightforward approaches steps 1–2). N and X present a phrase with negative and unknown sentiment respectively (these phrases do not contain a problem description).

Step 2. This step determines type of each sentence using a rule-based approach. The following rules cover three types of sentences:

(1) N | X, but IP | DP → PS (Type-A)
(2) IP, but X → ¬PS (Type-B)
(3) IP, but ¬DP → ¬PS (Type-B)
(4) ¬DP, but IP | N → ¬PS (Type-B)
(5) N | X, but N | X → ¬PS (Type-C)
(6) IP | DP, but IP | DP → PS (Type-C)

where PS is the sentence which the approach classifies as a problem sentence, while ¬PS is the sentence without problems.

Further, we compare the straightforward approach from Sect. 2.1 to the improved clause-based approach described in Sect. 2.1.

3   Evaluation and Experiments

For our experiments, we collected 1,496 sentences from the HP website.[3] To label the test set, we made use of the Amazon Mechanical Turk (MTurk) service. The MTurk task was to assign each sentence one of three following labels: (i) "a problem is indicated in text", (ii) "a problem is implicit" and (iii) "no problem in text". We treat both (i) and (ii) as problem labels. All other sentences were

---

[3] http://reviews.shop.hp.com

Table 1. Distribution of positive and negative labels after four MTurk runs

Subset	Positive marks	Class label	Sentences	Sentences with "but"
S0	0	-	562	134
S1	1	-	186	74
S2	2	unknown	145	83
S3	3	+	237	149
S4	4	+	366	201

Table 2. Performance metrics of straightforward and clause-based approaches

Method name	TPR	FPR	P	R	F1
Straightforward approach (Sect. 2.1)	.81	.27	.71	.81	.756
Clause-based approach (Sect. 2.2)	.77	.22	.74	.77	.754

added to no-problem class. We carried four separate runs with the same dataset in each run. We collected number of problem labels for each sentence after four runs. A total of 145 sentences were assigned two positive and two negative labels, so we excluded them from the test set. The distribution of labels in the evaluation corpus is presented in Table 1.

We analyzed number of sentences with but-conjunctions in the dataset. A total of 41 % of all sentences are composite sentences with *but*. The dataset we use in the evaluation contains 1,351 instances: 748 instances belong to the no-problem class (S3 + S4), and 603 instances belong to the problem class (S0 + S1). Performance metrics are calculated with this dataset and provided in Table 2.

We also compare our method to four machine learning methods from *Weka* toolkit.[4] The dataset we use for the classification contains 1,351 instances. Before applying any algorithm, we convert the data file into the attribute-relation file format (ARRF). Once data is transformed and loaded into *Weka*, we use bag-of-words model for representing sentences as collections of features. The StringToWordVector filter converts unstructured documents into feature vectors, keeping the frequency of each word.

First, we performed a tenfold cross validation using support vector machines (LibSVM scheme in *Weka*). The SVM classifier is a strong baseline used in sentiment analysis tasks. The experiment gave a precision of 88 %, a recall of 13 % and an F1 measure of 40 % on the dataset. We use a filter approach for feature selection to improve machine learning results. The AttributeSelection filter first orders features by their discriminative capacity (using the information gain), then removes unnecessary features for definition of a problem class.

We performed a tenfold cross validation on the dataset using the NaiveBayes-Multinomial, ZeroR, J48, and SMO schemes. We used sequential minimal optimization (SMO) for training a support vector classifier and Multinomial Naive

---

[4] http://www.cs.waikato.ac.nz/ml/weka/

Table 3. Comparison of machine learning algorithms and dictionary-based approaches.

	J48	NaiveBayes	SVM	Approach 2.1	Approach 2.2
Accuracy	.64	.65	.70	.77	**.78***
Precision	.61	.64	.66	.71	**.74***
Recall	.57	.60	.67	**.81***	.77
F1 measure	.59	.62	.67	**.76***	.75

* indicates that a metric result is statistically better than others in the same row

Bayes as a version of Naive Bayes for text documents represented by feature vectors. The percentage of correctly labeled instances for each of the four schemes is as follows: 55.3 % for majority-class baseline (ZeroR), 64.34 % for Decision Trees (J48), 66.94 % for Naïve Bayes (NaiveBayesMultinomial), 69.55 % for support vector classifier (SMO). The results of both SVM classifier and Naïve Bayes are statistically better than the baseline established by ZeroR and J48. The experiment shows that the most effective classifier among these methods is linear. A comparison of metrics without weighting (for a problem sentences class) is presented in Table 3. The straightforward approach (Sect 2.1) based on dictionaries is better than an SVM classifier on the dataset. The clause-based approach is also better than a machine-learning baseline.

4 Related Work

Problem-phrase extraction from user reviews of products has been mentioned in other papers. Gupta analyzed several hundred messages from Twitter and created dictionaries manually. He described several types of syntactic features: "verb particle" feature, "not-verb" feature, "verb-down" feature, and about 20 single verbs and nouns. Gupta [3] reported that performance of F1 measure was 0.66. In addition, Gupta used all the sentiment features and syntactic features. The best F1 measure was 0.742. In our approach, we have also used dictionaries to compile problem words. Our performance metrics are similar to those that have been shown in Gupta's [3,4], despite the difference in datasets and approaches.

But-conjuction has been widely studied in literature. In [2] Lakoff claims that *but* encodes a "denial-of-expectation" meaning between two conjuncts or "semantic opposition". In [5] Winter and Rimon assume that words like *but*, (*al*)*though, yet, nevertheless* are some of the common expressions in English. There are some strong relations between senses of *but*. Winter and Rimon developed an "application for analyzing the semantics and pragmatics of contrastive conjunctions in natural language". They showed that the case of *but* is different from those of *yet, although, nevertheless, even though* and other connectives of contrast. In [5] they introduce an intuitive definition for the connection between contrast and implication. Winter and Rimon use word implication to capture the notion of contrast. In our approach, we first look for words from the dictionaries in both clauses, then assign a label to each clause depending on the words with certain meanings. Other authors use the clause-based approach to solve text categorization tasks. In [6] Carreras and Marquez present a clause-based boosting approach for anti spam e-mail filtering. Their approach consists of decomposing the clause, splitting the problem into a combination of binary "simple" decision. They decompose a message in two levels with two chained decisions. At the first level, all clause candidates get a confidence value. The second level corresponds to the analysis of results of the first level by confidence values. Carreras and Marquez use open baskets to identify clauses, while we use but-conjunctions that introduce dependent clauses in a sentence.

5   Conclusion

In this paper, we propose using the clause-based approach to detect problem sentences in customer reviews. But-conjunctions in English give different meanings of contrast in different contexts. We decompose each sentence with a but-conjunction into the clauses, with separate analyses, and use a combination step to decide whether a sentence contains a problem or not. The clause-based approach is more effective for improving the precision of the classification; for high recall we recommend the straightforward dictionary-based approach. Both approaches perform very good compared to the simple baseline given by supervised machine-learning algorithms. In our future work, we plan to extract problem phrases with related targets for the Russian language. We plan to compare our approaches with a powerful baseline, based on conditional random fields (CRF), which have performed well in our previous work [7]. We plan to extend the dictionaries automatically to improve classification performance.

Acknowledgements. We are grateful to Valery Solovyev and Sergey Serebryakov for their support of this research, useful discussions and help with our approaches. We are grateful to the Program Committee members who provided constructive review comments. This work was partially supported by Russian Ministry of Education and Science (project number: 3056, "Semantic web technologies and linguistic databases: annotation, information extraction and retrieval").

References

1. Liu, B., Hu, M.: Mining and summarizing customer reviews. In: Proceedings of the Tenth ACM SIGKDD International Conference on Knowledge Discovery and Data Mining, KDD '04, New York, NY, USA (2004)
2. Lakoff, R.: If's, and's and but's about conjunction. In: Fillmore, L. (eds.) Studies in Linguistic Semantics, New York (1971)
3. Gupta, N.: Extracting descriptions of problems with product and service from twitter data. In: Proceedings of the 3rd Workshop on Social Web Search and Mining, Beijing, China, July 2011

4. Gupta, N.: Extracting phrases describing problems with products and services from twitter messages. Technical report, Conference on Intelligent Text Processing and Computational Linguistics CICling 2013, March 2013
5. Winter, Y., Rimon, M.: Contrast and implication in natural language. J. Semant. **11**, 365–406 (1994)
6. Carreras, X., Màrquez, L.: Boosting trees for clause splitting. In: Proceedings of the CoNLL-2001 Shared Task, Toulouse, France (2001)
7. Gareev, R., Tkachenko, M., Solovyev, V., Simanovsky, A., Ivanov, V.: Introducing baselines for Russian named entity recognition. In: Gelbukh, A. (ed.) CICLing 2013, Part I. LNCS, vol. 7816, pp. 329–342. Springer, Heidelberg (2013)

VKF-Method of Hypotheses Generation

Dmitry V. Vinogradov[1,2](✉)

[1] All-Russia Institute for Scientific and Technical Information (VINITI),
Intelligent Information Systems Laboratory, Moscow 125190, Russia
[2] Intelligent Robotics Laboratory, Russian State University for Humanities,
Moscow 125993, Russia
vin@viniti.ru
http://isdwiki.rsuh.ru

Abstract. We present an intelligent system for VKF-method based on
a Markov chain approach to generation of hypotheses about causes of
presence/absence of effect under study. The system uses coupling Markov
chains that terminate with probability 1. Since each hypothesis is generated by an independent run of the Markov chain, the system makes the
induction step in parallel by several threads. After that the abduction
step refines the hypotheses by the CloseByOne operation with training examples (in several threads too). Then the system predicts pres-
ence/absence of the effect by the analogical reasoning. We test the system
on SPECT dataset from UCI machine learning repository. The accuracy
is 85.56 percent (that exceeds 84.0 percent accuracy of the CLIP3 algorithm developed by the authors of the dataset).

Keywords: Machine learning · Markov chain · Termination with
probability 1 · Formal Concept Analysis · CloseByOne operation

1 Introduction

In 1983 Prof. Victor K. Finn [4] proposed many-valued logics formalization of
inductive methods of John Stuart Mill (JSM-method). These ideas underlie a
large class of intelligent systems of JSM type. JSM-method combines several
cognitive procedures: inductive generation of hypotheses from a training sample,
prediction of properties of test examples using analogy to the training ones,
abductive explanation of properties of training examples [5]. The state of the
art for JSM-method is represented in [6,7].

Initially, JSM community had developed original algorithms for the basic
procedures of JSM-method. Later Prof. Sergei O. Kuznetsov [15] discovered that
Formal Concept Analysis (FCA) [9] corresponds to the generation ("induction")
step of JSM-method. This relation allows us to use FCA algorithms for JSM-
method. For instance, modern JSM systems use the well-known algorithm of
Norris. Also there is an influence of JSM community on FCA. For example, the
well-known algorithm "Close-by-One" (CbO) was initially introduced by Prof.
Kuznetsov in [14] for JSM-method and later translated in terms of FCA.

© Springer International Publishing Switzerland 2014
D.I. Ignatov et al. (Eds.): AIST 2014, CCIS 436, pp. 237–248, 2014.
DOI: 10.1007/978-3-319-12580-0_25

In our opinion, the main drawback of 'old-fashioned' JSM-method is the computational complexity of JSM algorithms, especially for the induction step. Prof. Kuznetsov [16] provided theoretical bounds on computational complexities of various JSM related problems. Kuznetsov and Objedkov [17] presented results of experimental comparison between various (partially improved by the survey's authors) variants of famous deterministic algorithms of FCA.

The development of JSM-method has resulted in intelligent systems that were applied in various domains such as sociology [8,12], pharmacology, medicine [18], graphology [10,11], etc. In practice there were situations when a JSM system generated more than 10,000 formal concepts (JSM similarities) from a context with about 100 objects. In our opinion the importance of all generated concepts is doubtful, because when domain experts manually select important JSM causes they reject majority of generated JSM similarities.

In the paper [19] we introduce the Markov chain approach (see [3], for instance) to the generation of JSM hypotheses. The paper [20] describes this approach in FCA terms. Here we extend it to a full-scale machine learning model called VKF-method. To do this the abduction rule is replaced by the procedure of abductive refinement of hypotheses with respect to the training examples.

On the basis of VKF-method scheme we create an intelligent system that corresponds to the simplest JSM strategy. The system uses the coupling Markov chain to generate random sample of hypotheses. Each run of this chain terminates with probability 1. Since each hypothesis is generated by an independent run of the Markov chain, the system makes the induction step in parallel by several threads. Then the abduction step refines the hypotheses by the CloseByOne operation with training examples again in several threads. Finally, the system predicts target class of each test example by the analogy reasoning.

We test the system on the dataset SPECT from machine learning repository at University California Irvine [1]. The accuracy of prediction of the system is equal to 85.56 percent (which exceeds 84.0 percent accuracy of the CLIP3 algorithm [13] developed by the authors of the dataset).

2   Background

2.1   Basic Definitions and Facts of FCA

Here we recall some basic definitions and facts of Formal Concept Analysis (FCA).

A **(finite) context** is a triple (G, M, I), where G and M are finite sets and $I \subseteq G \times M$. The elements of G and M are called **objects** and **attributes**, respectively. Conventionally, we write gIm instead of $\langle g, m \rangle \in I$ to denote that object g has attribute m.

For $A \subseteq G$ and $B \subseteq M$, define

$$A' = \{m \in M \mid \text{for all } g \in A(gIm)\}, \tag{1}$$

$$B' = \{g \in G \mid \text{for all } m \in B(gIm)\}; \tag{2}$$

so A' is the set of all attributes common to all objects in A and B' is the set of all objects having all attributes in B. The maps $(\cdot)' : A \mapsto A'$ and $(\cdot)' : B \mapsto B'$ are called **derivation operators** (**polars**) of the context (G, M, I).

A **concept** of the context (G, M, I) is a pair (A, B), where $A \subseteq G$, $B \subseteq M$, $A' = B$, and $B' = A$. The first component A of the concept (A, B) is called the **extent** of the concept, and the second component B is called its **intent**. The set of all concepts of the context (G, M, I) is denoted by $\mathbf{B}(G, M, I)$.

Let (G, M, I) be a context. For concepts (A_1, B_1) and (A_2, B_2) in $\mathbf{B}(G, M, I)$ we write $(A_1, B_1) \leq (A_2, B_2)$, if $A_1 \subseteq A_2$. The relation $\leq$ is a **partial order** on $\mathbf{B}(G, M, I)$.

A subset $A \subseteq G$ is the extent of some concept if and only if $A'' = A$ in which case the unique concept with extent A is (A, A'). Similarly, a subset $B \subseteq M$ is the intent of some concept if and only if $B'' = B$ and thus the unique concept with the intent B is (B', B).

It is easy to check that $A_1 \subseteq A_2$ implies $A_1' \supseteq A_2'$ and for concepts (A_1, A_1') and (A_2, A_2') reverse implication is valid too, because $A_1 = A_1'' \subseteq A_2'' = A_2$. Hence, for (A_1, B_1) and (A_2, B_2) in $\mathbf{B}(G, M, I)$

$$(A_1, B_1) \leq (A_2, B_2) \Leftrightarrow A_1 \subseteq A_2 \Leftrightarrow B_2 \subseteq B_1. \tag{3}$$

Let us consider a context (G, M, I). In what follows, let J be an index set. We assume that $A_j \subseteq G$ and $B_j \subseteq M$, for all $j \in J$.

Lemma 1. *[9] Assume that (G, M, I) is a context and let $A \subseteq G$, $B \subseteq M$ and $A_j \subseteq G$ and $B_j \subseteq M$ for all $j \in J$. Then*

$$A \subseteq A'' \quad and \quad B \subseteq B'', \tag{4}$$

$$A_1 \subseteq A_2 \Rightarrow A_1' \supseteq A_2' \quad and \quad B_1 \subseteq B_2 \Rightarrow B_1' \supseteq B_2', \tag{5}$$

$$A' = A''' \quad and \quad B' = B''', \tag{6}$$

$$\left(\bigcup_{j \in J} A_j\right)' = \bigcap_{j \in J} A_j' \quad and \quad \left(\bigcup_{j \in J} B_j\right)' = \bigcap_{j \in J} B_j', \tag{7}$$

$$A \subseteq B' \quad \Leftrightarrow \quad A' \supseteq B. \tag{8}$$

Proposition 1. *[9] Let (G, M, I) be a context. Then $(\mathbf{B}(G, M, I), \leq)$ is a lattice with join and meet operators given by*

$$\bigvee_{j \in J} (A_j, B_j) = \left(\left(\bigcup_{j \in J} A_j\right)'', \bigcap_{j \in J} B_j\right), \tag{9}$$

$$\bigwedge_{j \in J} (A_j, B_j) = \left(\bigcap_{j \in J} A_j, \left(\bigcup_{j \in J} B_j\right)''\right). \tag{10}$$

Corollary 1. *For context (G, M, I) the lattice $(\mathbf{B}(G, M, I), \leq)$ has (M', M) as the bottom element and (G, G') as the top element. In other words, for all $(A, B) \in \mathbf{B}(G, M, I)$ the following inequalities hold:*

$$(M', M) \leq (A, B) \leq (G, G'). \tag{11}$$

2.2   The "Close-by-One" Operations: Definition and Properties

By means of the infimum and supremum definitions in $\mathbf{B}(G, M, I)$ given by Proposition 1 we can introduce local steps of our Markov chains:

Definition 1. *For $(A, B) \in \mathbf{B}(G, M, I)$, $g \in G$, and $m \in M$ define*

$$CbO((A, B), g) = ((A \cup \{g\})'', B \cap \{g\}'), \tag{12}$$
$$CbO((A, B), m) = (A \cap \{m\}', (B \cup \{m\})''). \tag{13}$$

so $CbO((A, B), g)$ is equal to $(A, B) \vee (\{g\}'', \{g\}')$ and $CbO((A, B), m)$ is equal to $(A, B) \wedge (\{m\}', \{m\}'')$.

We call these operations CbO because the first one is used in Close-by-One (CbO) algorithm to generate all concepts from $\mathbf{B}(G, M, I)$, see [14] for details.

Lemma 2. *Let (G, M, I) be a context, $(A, B) \in \mathbf{B}(G, M, I)$, $g \in G$, and $m \in M$. Then*

$$g \in A \Rightarrow CbO((A, B), g) = (A, B), \tag{14}$$
$$m \in B \Rightarrow CbO((A, B), m) = (A, B), \tag{15}$$
$$g \notin A \Rightarrow (A, B) < CbO((A, B), g), \tag{16}$$
$$m \notin B \Rightarrow CbO((A, B), m) < (A, B). \tag{17}$$

Proof. If $g \notin A$ then $A \subset A \cup \{g\} \subseteq (A \cup \{g\})''$ by (5). By definition of the order on concepts this inclusion and (12) imply (16). Implication (17) is proved in the same way, the rest is obvious.

Lemma 3. *Let (G, M, I) be a context, $(A_1, B_1), (A_2, B_2) \in \mathbf{B}(G, M, I)$, $g \in G$, and $m \in M$. Then*

$$(A_1, B_1) \leq (A_2, B_2) \Rightarrow CbO((A_1, B_1), g) \leq CbO((A_2, B_2), g), \tag{18}$$
$$(A_1, B_1) \leq (A_2, B_2) \Rightarrow CbO((A_1, B_1), m) \leq CbO((A_2, B_2), m). \tag{19}$$

Proof. If $A_1 \subseteq A_2$ then $A_1 \cup \{g\} \subseteq A_2 \cup \{g\}$. Hence (5) implies $(A_2 \cup \{g\})' \subseteq (A_1 \cup \{g\})'$. Second part of (5) implies $(A_1 \cup \{g\})'' \subseteq (A_2 \cup \{g\})''$. By definition of the order between concepts this is (18). Implication (19) is proved in the same way by using (3).

2.3   Example of JSM-Reasoning

We represent JSM-method by its application to a school problem: find sufficient conditions for a convex quadrangle with symmetries to be circled and (using these conditions) tell whether a rectangle can be circled. Hence there are two target classes: positive (there exists a circle around a quadrangle) and negative (otherwise).

The training sample contains a square, an isosceles trapezoid, a diamond, and a deltoid (see the rows labels in Table below). The test example is rectangle.

We represent each quadrangle by the subset of attributes related to its possible symmetries: "There exists a central symmetry point" (A), "The group of rotations is trivial" (B), "The group of rotations contains at least two elements" (C), "There is a diagonal symmetry axis" (D), "There is a non-diagonal symmetry axis" (E). See the columns labels in Table below.

quadrangle	target	A	B	C	D	E
square	1	1	0	1	1	1
trapezoid	1	0	1	0	0	1
diamond	0	1	0	1	1	0
deltoid	0	0	1	0	1	0
rectangle	?	1	0	1	0	1

The **induction step** of JSM-method is a procedure for separately finding all concepts of the positive part of the training sample and all concepts of the negative part.

The positive context generates 4 concepts:

extent	A	B	C	D	E
$\emptyset$	1	1	1	1	1
$\{square\}$	1	0	1	1	1
$\{trapezoid\}$	0	1	0	0	1
$\{square, trapezoid\}$	0	0	0	0	1

Then JSM-system tests the **counter-example forbidding condition**: is there a negative example that includes a positive concept intent? JSM-system saves only concepts that do not have any counter-examples. Also there is a lower bound on the minimal number of examples in concept extents (**its parents list**). Usually, this bound is set to 2. Hence the JSM-system saves only the last concept ($\{square, trapezoid\}, \{E\}$) as a JSM-hypothesis.

The **abduction step** of JSM-method is a procedure, which "explains" all the training examples by saved JSM-hypotheses.

Since the saved JSM-hypothesis ($\{square, trapezoid\}, \{E\}$) is included $\{E\} \subseteq \{A, C, D, E\}$ and $\{E\} \subseteq \{B, E\}$ into both positive examples, the system explains them by this hypothesis.

The **analogy step** of JSM-method is a procedure to predict the target class of test examples. Since the saved JSM-hypothesis ($\{square, trapezoid\}, \{E\}$) is contained $\{E\} \subseteq \{A, C, E\}$ in the test example ($\{rectangle\}, \{A, C, E\}$), the system predicts that there is a circle around rectangle.

3 Markov Chain Algorithm

Now we describe coupling Markov chain algorithm for random generation of formal concepts.

Let S be a finite set (**state space**). A **Markov chain** is defined by a (stochastic) matrix $K(x, y)$ with $\forall x \in S \forall y \in S[K(x, y) \geq 0]$ and $\Sigma_{y \in S} K(x, y) = 1$ for each $x \in S$. Thus each row is a probability measure on S, so K defines a kind of random walk: from $x \in S$ choose $y \in S$ with probability $K(x, y)$; from y choose $z \in S$ with probability $K(y, z)$, and so on.

We refer to the sequence of states $X_0 = x, X_1 = y, X_2 = z, \ldots$ as a **run of the chain** starting at x. From the definitions $P(X_1 = y \mid X_0 = x) = K(x, y)$, $P(X_2 = z, X_1 = y \mid X_0 = x) = K(x, y) \cdot K(y, z)$.

It implies

$$P(X_2 = z \mid X_0 = x) = \Sigma_{y \in S} K(x, y) \cdot K(y, z) = K^2(x, z). \tag{20}$$

Similarly, the nth power of the matrix K has in x, y entry

$$K^n(x, y) = P(X_n = y \mid X_0 = x). \tag{21}$$

Now we represent the coupling Markov chain algorithm, which is the core of VKF-method.

Data: context (G, M, I), external function $CbO(\ ,\ )$
Result: random concept $(A, B) \in \mathbf{B}(G, M, I)$
$X := G \sqcup M$; $(A, B) := (M', M)$; $(C, D) = (G, G')$;
while $((A \neq C) \vee (B \neq D))$ **do**
  select random element $x \in X$;
  $(A, B) := CbO((A, B), x)$; $(C, D) := CbO((C, D), x)$;
end

Algorithm 1. Coupling Markov chain

The intermediate value of quadruple $(A, B) \leq (C, D)$ at some step t of the while loop of Algorithm 1 corresponds to Markov chain state X_t. The A_t, B_t, C_t and D_t are first, second, third, and fourth components, respectively.

If we compute all the concepts for context (G, M, I), then it is possible to give the exact matrix representation of our Markov chain. However, an advantage of this approach is that one does not need to know the transition matrix for the algorithm to run. Our goal is to select a random sample of concepts without computing (possibly exponentially large) set $\mathbf{B}(G, M, I)$ of all concepts. Why does the algorithm correspond to a Markov chain? The CbO operation with respect to some fixed object defines a deterministic function (non-permutation, in general) from $\mathbf{B}(G, M, I)$ into itself. For every row the corresponding matrix contains a single 1 in some cell and 0's in other places. So the matrix is a stochastic one. The similar fact holds for operation CbO with respect to a fixed attribute.

Because the transition for elements of $\mathbf{B}(G, M, I)$ is defined by the random (uniform) choice of either object or attribute, the transition matrix is the (uniformly) weighted sum of the corresponding matrices for every object and every attribute. However, the weighted sum of stochastic matrices is a stochastic matrix. Thus the transition matrix for elements of $\mathbf{B}(G, M, I)$ is stochastic one.

Simultaneous application of CbO operation with respect to same either object or attribute to an ordered pair of elements of $\mathbf{B}(G, M, I)$ corresponds to tensor (Kronecker) product of the transition matrix on $\mathbf{B}(G, M, I)$ by itself with respect to the invariant subspace of ordered pairs. The invariance of this subspace is proved in Lemma 3. The tensor product of two stochastic matrices is a stochastic matrix. The restriction of a stochastic matrix on invariant subspace is a stochastic matrix too. Hence the transition matrix for Algorithm 1 is stochastic one and determines a Markov chain.

Definition 2. *A **coupling length** for context (G, M, I) is defined by*

$$L = \min(\mid G \mid, \mid M \mid). \tag{22}$$

*A **choice probability** for an object or an attribute in context (G, M, I) is equal to*

$$p = \frac{1}{\mid G \mid + \mid M \mid}. \tag{23}$$

Lemma 4. *If $\mid G \mid \geq \mid M \mid$, then for every integer r and every initial state $(A, B) \leq (C, D)$ (where $(A, B), (C, D) \in \mathbf{B}(G, M, I)$) the following holds:*

$$\mathrm{P}[A_{r+L} = C_{r+L} = A, B_{r+L} = D_{r+L} = B \mid Y_r = (A, B) \leq (C, D)] \geq p^L. \tag{24}$$

If $\mid G \mid \leq \mid M \mid$ then for every integer r and every initial state $(A, B) \leq (C, D)$ (where $(A, B), (C, D) \in \mathbf{B}(G, M, I)$) the following holds

$$\mathrm{P}[A_{r+L} = C_{r+L} = C, B_{r+L} = D_{r+L} = D \mid Y_r = (A, B) \leq (C, D)] \geq p^L. \tag{25}$$

Proof. For coupling to $Y_{r+L} = (A, B) \leq (A, B)$ one needs to move the upper part $(C, D) \in \mathbf{B}(G, M, I)$ of the Markov chain state $Y_r = (A, B) \leq (C, D)$ to (A, B) when the lower part (A, B) is fixed. To this end the algorithm can enumerate all elements of $B \setminus D$ in a fixed order and then apply CbO operator to this sequence. However $\mid B \setminus D \mid \leq \mid M \mid = L$. Relation (25) is proved in a similar way. □

Remark 1. The previous proof produces a very weak lower bound on the probability of the coupling. We can strengthen this result if we consider all possible ways from the upper element (C, D) to the lower one (A, B). Then the bound $(\mid G \mid + \mid M \mid)^{-\mid M \mid}$ is replaced by $\mid M \mid! / (\mid G \mid + \mid M \mid)^{\mid M \mid}$.

Example 1. Let us consider a finite set G with $n = \mid G \mid$ elements. The simplest example of a concepts lattice is $\mathbf{B}(G, G, \neq)$, where $\neq = \{(g, m) \in G \times G \mid g \neq m\}$. The lattice $\mathbf{B}(G, G, \neq)$ is isomorphic to $\{A \mid A \subseteq G\}$ with respect to the set operations $\bigcap$ and $\bigcup$. In this example the coupling length is equal to n and the choice probability is equal to $1/(2n)$.

Remark 2. Let us compare two lower bounds from lemma 4 and Remark 1 for the case of $\mathbf{B}(G, G, \neq)$. Lemma 4 produces $(2n)^{-n}$ and Remark 1 makes $n! / (2n)^n$. Stirling's approximation converts the last bound into $\sqrt{2\pi n} / (2e)^n$. The improvement is essential because the last bound is only exponentially small.

Remark 3. For Example 1 we can improve the bound further to $\sqrt{2\pi n}/e^n$. It needs to consider a possibility of coupling for any concept $(E, F) \in \mathbf{B}(G, G, \neq)$ with $(A, B) \leq (E, F) \leq (C, D)$.

Theorem 1. *The coupling Markov chain has the probability of coupling (termination) before n steps with limit 1 as $n \to \infty$.*

Proof. Lemma 3 implies that $(A_{(k-1)\cdot L}, B_{(k-1)\cdot L}) \leq (C_{(k-1)\cdot L}, D_{(k-1)\cdot L})$. Let $r = (k-1) \cdot L$. Then Lemma 4 implies that $\mathrm{P}[A_{k \cdot L} \neq C_{k \cdot L} \vee B_{k \cdot L} \neq D_{k \cdot L} | Y_r = (A_r, B_r)] \leq (1 - p^L)$. After k independent repetitions we have $\mathrm{P}[A_{k \cdot L} \neq C_{k \cdot L} \vee B_{k \cdot L} \neq D_{k \cdot L} | Y_0 = (M', M) \leq (G, G')] \leq (1 - p^L)^k$. However when $k \to \infty$ we have $(1 - p^L)^k \to 0$. $\qquad\square$

Remark 4. The case $\mathbf{B}(G, G, \neq)$ admits a direct computation of the mean time to couple. The result is $O(n \cdot \ln n)$. However there are difficulties to obtain a similar result for the general case of the concept lattice of an arbitrary context.

4   The Simplest Program for VKF-method

4.1   General Structure of the Simplest Variant of VKF-method

Now we represent the general scheme of the simplest program based on VKF-method.

At first, the program reads two files with training and test examples, respectively. Each line of these files corresponds to a single example. The first field (position) of a line represents the target class of the corresponding example. The next fields (separated by commas) represent binary attributes of the corresponding example. We assume that there are two target classes. There are standard (from JSM-method's point of view) extensions of VKF-method on target classes that form a lower semilattice.

Starting from positive examples of the training sample the program generates a formal context (G^+, M, I). The negative examples form the list G^- of obstacles (counterexamples in JSM terms). The program also creates the list G^τ of examples to predict the target class from all the test examples.

After that the program apply the coupling Markov chain algorithm to generate a random formal concept $(A, B) \in \mathbf{B}(G^+, M, I)$. The program saves the concept (A, B) if there is no obstacle $S \in G^-$ such that $B \subseteq S$ and $|A| \geq 2$. This procedure corresponds to **the induction step of JSM-method**. We replace a time-consuming deterministic algorithm (for instance, "Close-by-One") for generation of all concepts by the probabilistic one to randomly generate the prescribed number of concepts.

Then the program applies a modification of **the abduction step of JSM-method**. Since the program generates only a (possibly small) fraction of hypotheses, it refines generated hypotheses by applying CloseByOne operation with respect to all positive examples. Again VKF-program tests absence of obstacles for new hypotheses (**the counter-example forbidding condition** in JSM terms). If at least one of newly generated hypotheses satisfies this condition the corresponding example is declared to be explained by this hypothesis. Otherwise the example is unique and the program shows it to a domain expert to extend the training sample by similar examples.

Finally, the program predicts the target class of the test examples and compares the prediction results with the original target value. This procedure coincides with the **the analogy step of JSM-method.**

4.2 Implementation Details of the Simplest Program

The simplest variant of VKF-method was implemented as a C++ program using the boost library. The examples are represented by objects of *boost :: dynamic_bitset* class from the boost library. They are stored in containers of *std :: vector* and *std :: list* classes of the standard C++ library.

The program uses *boost :: random* classes for random number generator. It applies to the Markov chain step.

The multi-threading is implemented by POSIX pthread. There are *boost :: thread* classes but the POSIX pthread approach is faster and based on Windows' API for Microsoft Windows. However the program is platform-independent.

The program is created as a console application. We use open-source C++ IDE Code::Blocks (version 13.12) together with the latest boost library (version 1_56_0). The C++ compiler is GNU C++ toolset (version 4.9.1).

5 Experimental Validation of the Approach

The VKF-method system was applied to SPECT dataset [13] from the machine learning repository of the University of California in Irvine (UCI) [1]. The training set contains 40 positive and 40 negative examples. The test sample contains 172 positive and 15 negative examples. Each example is described by 22 initial binary attributes. This set of attributes was updated by attributes staying for negations of the initial attributes, so the positive context was a 40×44 matrix.

Since each hypothesis is generated through independent run of the coupling Markov chain algorithm, the VKF-method program used several threads to compute the induction step. On 4 threads CPU (i5-3210M) the maximal CPU's load for 4 threads computation attained 90 percent. The abduction step was computed in several threads too. The prediction of the target class of test examples was carried out in one thread because of lower computational complexity of the analogy step in comparison to the induction step.

The accuracy of predicting the target class of test examples was 85.56 percent (151 of 172 positive cases and 9 of 15 negative ones). The authors of SPECT dataset attained 84.0 percent accuracy [13] by own machine learning system CLIP3 that computes the cover by means of integer programming [2]. The authors of the dataset assert that the ensemble of CLIP4 classifiers attained 90.4 percent accuracy. However, the weighted nature of the ensemble approach is a serious drawback from the viewpoint of domain experts.

Data: files of training and test samples; the number N of hypotheses to generate.

Result: number of correctly predicted elements of test sample.

$G^+ :=$ positive training examples, $M :=$ binary attributes; $I \subseteq G^+ \times M$ is formal context corresponding to the positive part of training sample;

$(A_{max}, B_{max}) := (M', M)$; $(C_{min}, D_{min}) = (G^+, G'^+)$;

$G^- :=$ negative training examples;

$G^\tau :=$ elements of test sample;

$Hyp(G^+, M, I, G^-) := \emptyset$; $i := 0$;

while $(i < N)$ **do**

    select random concept $(A, B) \in \mathbf{B}(G^+, M, I)$ by means of the coupling Markov chain algorithm; $hasObstacle :=$ **false**;

    for $(g \in G^-)$ **do**

        if $(B \subseteq g')$ **then**

            $hasObstacle :=$ **true**;

        end

    end

    if $(hasObstacle = \textit{false})$ **then**

        $Hyp(G^+, M, I, G^-) := Hyp(G^+, M, I, G^-) \cup \{(A, B)\}$;

        $i := i + 1$;

    end

end

for $(g^+ \in G^+ \text{ and } (A, B) \in Hyp(G^+, M, I, G^-))$ **do**

    compute $(X, Y) = CbO((A, B), g^+)$;

    $Explained(g^+) :=$ **false**; $hasObstacle :=$ **false**;

    for $(g \in G^-)$ **do**

        if $(Y \subseteq g')$ **then**

            $hasObstacle :=$ **true**;

        end

    end

    if $(hasObstacle = \textit{false})$ **then**

        $Hyp(G^+, M, I, G^-) := Hyp(G^+, M, I, G^-) \cup \{(X, Y)\}$;

        $Explained(g^+) :=$ **true**;

    end

end

for $(g \in G^\tau \text{ and } (A, B) \in Hyp(G^+, M, I, G^-))$ **do**

    if $(B \subseteq g')$ **then**

        $PredictPositively(g) :=$ **true**;

    end

end

Algorithm 2. General scheme of VKF-method

6 Conclusions

In this paper we have described the VKF-method system of hypotheses generation by coupling Markov chain for random generation of concepts of a finite context. The coupling algorithm terminates with probability 1. We extend the abduction step by additional refinement procedure on the basis of CloseByOne operation.

We validate our approach by predicting target class of test examples of SPECT dataset from UCI machine learning repository [1]. Accuracy of the simplest model of VKF-method supersedes results of CLIP3 system proposed by the dataset's authors [13]. There are more sophisticated models of VKF-method corresponding to those of JSM-method. The study of these models will be the matter of further studie.

Acknowledgements. The author would like to thank Prof. Victor K. Finn, Prof. Sergei O. Kuznetsov, and Dr. Sci. Maria A. Mikheyenkova for helpful discussions on JSM-method and Tatyana A. Volkova for programming support. The author thanks the anonymous referees for the careful reviews and the valuable comments that helped to improve the representation. The research was supported by Russian Foundation for Basic Research (project 14-07-00856a) and Presidium of the Russian Academy of Science (Fundamental Research Program 2014-2015).

References

1. Bache, K., Lichman, M.: UCI Machine Learning Repository. SICS UCI, Irvine CA (2013). http://archive.ics.uci.edu/ml
2. Cios, K.J., Wedding, D.K., Liu, N.: CLIP3: cover learning using integer programming. Kybernetes **26**(4–5), 513–536 (1997)
3. Diaconis, P.: The Markov Chain Monte Carlo revolution. Bull. Am. Math. Soc. **46**(2), 179–205 (2009)
4. Finn, V.K.: About machine-oriented formalization of plausible reasonings in F. Beckon-J.S. Mill Style. Semiotika i Informatika (Russian) **20**, 35–101 (1983)
5. Finn, V.K.: The synthesis of cognitive procedures and the problem of induction. Autom. Doc. Math. Linguist. **43**, 149–195 (2009)
6. Finn, V.K.: J.S. Mill's inductive methods in artificial intelligence systems I. Sci. Tech. Inf. Process. **38**(6), 385–402 (2011)
7. Finn, V.K.: J.S. Mill's inductive methods in artificial intelligence systems II. Sci. Tech. Inf. Process. **39**(5), 241–260 (2012)
8. Finn, V.K., Mikheyenkova, M.A.: Plausible reasoning for the problems of cognitive sociology. Log. Log. Phylos. **20**, 113–139 (2011)
9. Ganter, B., Wille, R.: Formal Concept Analysis: Mathematical Foundations. Springer, Heidelberg (1999)
10. Gusakova, S.M.: Similarity operation in the graphological examination identification problem. Nauch. Tekhn. Inf. Ser. 2 (Autom. Doc. Math. Linguist.) **44**(2), 64–67 (2010)
11. Gusakova, S.M., Komarov, A.S.: The possibility of the using jsm method for solving the problems of graphological expertise. Nauch. Tekhn. Inf. Ser. 2 (Autom. Doc. Math. Linguist.) **41**(5), 209–216 (2007)
12. Klimova, S.G., Mikheyenkova, M.A.: Formal methods of situational analysis: experience from their use. Nauch. Tekhn. Inf. Ser. 2 (Autom. Doc. Math. Linguist.) **46**(4), 183–194 (2012)
13. Kurgan, L.A., Cios, K.J., Tadeusiewicz, R., Ogiela, M., Goodenday, L.S.: Knowledge discovery approach to automated cardiac SPECT diagnosis. Artif. Intell. Med. **23**(2), 149–169 (2001)

14. Kuznetsov, S.O.: A fast algorithm for computing all intersections of objects in a finite semi-lattice. Nauch. Tekhn. Inf. Ser. 2 (Autom. Doc. Math. Linguist.) **27**(5), 11–21 (1993)
15. Kuznetsov, S.O.: Mathematical aspects of concept analysis. J. Math. Sci. **80**(2), 1654–1698 (1996)
16. Kuznetsov, S.O.: Complexity of learning in concept lattices from positive and negative examples. Discrete Appl. Math. **142**(1–3), 111–125 (2004)
17. Kuznetsov, S.O., Obiedkov, S.A.: Comparing performance of algorithms for generating concept lattices. J. Exp. Theor. Artif. Intell. **14**(2–3), 189–216 (2002)
18. Pankratova, E.S.: JSM-type intelligent systems for clinical data analysis. Autom. Doc. Math. Linguist. **45**(2), 81–90 (2011)
19. Vinogradov, D.V.: Random Generation of hypotheses in the JSM method using simple Markov Chains. Autom. Doc. Math. Linguist. **46**(5), 221–228 (2012)
20. Vinogradov, D.V.: A Markov Chain approach to random generation of formal concepts. In: Carpineto, C., Kuznetsov, S.O., Napoli, A. (eds.) FCAIR 2013, pp. 127–133. HSE, Moscow (2013)

Neural Models for Recognition
of Basic Units of Semiographic Chants

Ekaterina Vylomova[1,2]([✉]), Andrey Philippovich[3], Marina Danshina[3],
Irina Golubeva[3], and Yuriy Philippovich[3]

[1] Moscow University of Printing Arts, Moscow, Russia
evylomova@it-claim.ru
[2] Montclair State University, Montclair, USA
[3] Bauman Moscow State Technical University, Moscow, Russia
{aphilippovich,mdanshina,igolubeva,y_philippovich}@it-claim.ru

Abstract. The paper presents research on the problem of pattern recognition applied to the analysis of basic units of ancient Russian chants. We take for testing two types of neural networks: Multilayer Perceptron (MLP) and Probabilistic Neural Network (PNN). We investigate main features of the chant units and the properties of the networks to choose the best structure and algorithm. The results provide an analysis of accuracy for both approaches used in solving this particular task.

Keywords: Pattern recognition · Neural networks · Semiography

1 Introduction

One of the most important and prominent tasks in the study of ancient Russian culture is the exploration of melodic content in vocal music manuscripts from the XII–XVII centuries. The melodies in these books are written using special musical structures that evolved in Russia over the centuries. The concept «semiography» is understood to mean the conventionally accepted methods of musical writing and expression of certain musical sounds and how they are related. Figure 1 shows a fragment of a semiographic chant.

The structure of a chant consists of several parts:

- Flags, or *znamyas* (semiographic symbols) meaning musical symbols;
- Text matching flags;
- *Pometas* indicating the duration and amplitude of the music.

During 2000s these chants were gathered in one database, digitalized and a group of researchers [1–3] performed statistical analysis on these data to reveal and recover the knowledge which was forgotten or lost.

This paper presents a system that automates the recognition process for manuscripts of this kind. In particular, during the current stage we were able to devise a way to classify the basic parts of chants called *pometas*.

© Springer International Publishing Switzerland 2014
D.I. Ignatov et al. (Eds.): AIST 2014, CCIS 436, pp. 249–254, 2014.
DOI: 10.1007/978-3-319-12580-0_26

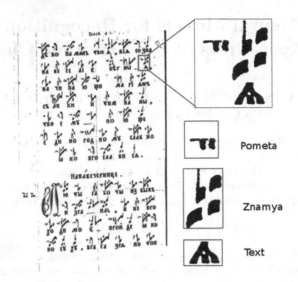

Fig. 1. A semiographic chant

2   Materials and Methods

2.1   Dataset

During the research we considered 7 most common types of *pometas* (see Table 1). The initial set contained 429 examples. It was randomly divided into 2 separate sets: training and test. The training set had 70 % of the examples and the remaining 30 % formed test set.

The table below presents the number of examples of each type in the training and test sets.

Table 1. A list of *pometas*

Pometa	Training set	Test set	Usage frequency
"С"	97	28	0.25
"Р"	78	24	0.16
"Н"	68	24	0.21
"М"	65	20	0.25
"П"	62	20	0.12
"Г"	31	10	0.07
"В"	28	9	0.04

2.2   Feature Selection

Initially, we manually cropped the *pometas* from chant manuscripts. In the next step we removed the noise and extra white space from the images. Then we automatically extracted 10 geometrical features:

- Number of intersections in the horizontal plane, i.e. the number of times a horizontal line intersected the borders of a *pometa*. We compared the results for different number of lines (3, 5 and 7) and found out that 5 lines (see the figure below) provided better rate of classification.
- Number of intersections in the vertical plane, i.e. the number of times a vertical line intersected the borders of a *pometa*. Here we applied the same procedure as described above and 5 lines proved to provide a better rate of classification (Fig. 2).

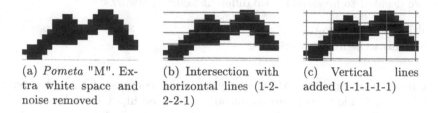

(a) *Pometa* "M". Extra white space and noise removed

(b) Intersection with horizontal lines (1-2-2-2-1)

(c) Vertical lines added (1-1-1-1-1)

Fig. 2. Extracting geometrical features

Figure 3a presents clusterization of resulting vectors in a 2-dimensional space.

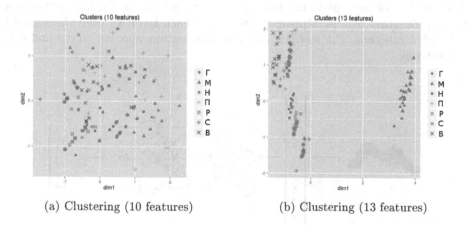

(a) Clustering (10 features)

(b) Clustering (13 features)

Fig. 3. Clustering results

We applied multidimensional scaling technique [4] to project 10 dimensions on the 2D plot. The classification would not be reliable due to mixed clusters, therefore we added extra features to improve the clusterization rate:

- Number of inclined lines used in the *pometa*.
- Number of horizontal lines used in the *pometa*.
- Number of vertical lines used in the *pometa*.

Figure 3b presents the results of new clusterization. Most of the clusters are highly separated. Only the clusters for *pometas* "Π" and "H" are somewhat mixed together.

Finally, each *pometa* is described as a vector with 13 dimensions (5 + 5 + 3). For example, a vector representation of the *pometa* "M" might be $x_i = (1; 2; 2; 2; 1; 1; 1; 1; 1; 1; 1; 4; 0; 0)$.

2.3   Classification

During the research we used statistics-based methods of classification. In particular, we decided to investigate two different neural networks:

- multilayer perceptron;
- probabilistic neural network.

Multilayer Perceptron. Multilayer perceptron consists of three layers: an input layer, a hidden layer and an output layer. The input layer has 14 units, or neurons. The neurons correspond to a feature vector and 1 neuron is always set to 1 and used for regulation. Neurons of the output layer correspond to the binary code of a class. Each of 7 classes is binary encoded. The length H of binary code equals to the number of classes. The binary code has $(H - 1)$ values equal to 0 and single value equal to 1, that corresponds to the right classification. The number of neurons in the hidden layer was set to 15. Figure 4 presents a typical structure of a perceptron network. For each layer we used the sigmoid activation function. The learning rate and momentum were set to 0.9 and 0.1, respectively.

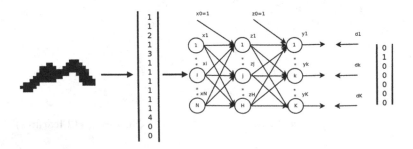

Fig. 4. Multilayer perceptron

Probabilistic Neural Networks. Probabilistic neural networks (PNN) [5] are derived from Bayesian networks and kernel Fisher discriminant analysis. PNNs have input, example, summation and output layers. In the current case the input layer corresponds to the feature vector and has 13 inputs. The example layer stores all possible examples of train data. The neurons of the summation layer are related to the categories (classes). In our case, the number of neurons in the example and summation layers equal to 294 and 7, respectively. The output layer has one decision neuron which chooses the most probable answer among all categories.

Figure 5 presents the structure of the probabilistic network we used to classify *pometas*.

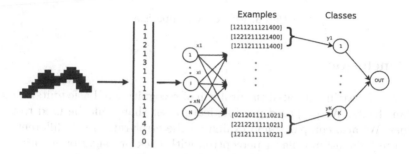

Fig. 5. A probabilistic neural network

Error Rate Evaluation. Error of the networks is evaluated as MSE: $Err = \frac{1}{2}\sum_{k=1}^{K}(d_k - y_k)^2$, where K is the number of examples, d_k is a desired output for kth example, y_k is actual output for kth example.

3 Results

During the experiments we got the following percentage of correct answers: $MLP_{train} = 0.96$, $MLP_{test} = 0.92$, $PNN_{train} = 0.96$, $PNN_{test} = 0.93$.

ROC curves on the figure below demonstrate that both classifiers experienced problems with differentiating *pometas* "Π" and "H". These *pometas* are written somewhat similar and it's more difficult to differentiate them than others. This could be predicted from the results of clustering during the feature selection stage.

According to the data, PNN network had better results and less error. Also, PNN is easier in usage and it does not need a lot of configuration and properly set network parameters such as activation functions, momentum and learning rates, number of iterations in case of perceptron. But we should also mention that in our study we had very few training examples, for large datasets PNN needs a lot of memory to store each example (Fig. 6).

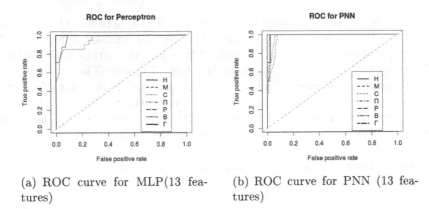

(a) ROC curve for MLP(13 features)   (b) ROC curve for PNN (13 features)

Fig. 6. Efficiency comparison

4   Conclusion

In this article, we investigated the problem of recognition of basic units of ancient Russian chants. We proposed possible feature set that could be used to train a classifier. We also compared and evaluated the error rates for 2 different classification techniques: multilayer perceptron with back-propagation algorithm and probabilistic neural network. We showed the benefits of the latter model, e.g. less error rate, less dependency on network settings set. Finally, we proved that PNNs exhibit better behaviour in pattern recognition tasks with few training examples.

Acknowledgements. The research is supported by grant 11-04-12025 of Russian Humanitarian Scientific Fund.

References

1. Golubeva, I.V., Yu, P.A.: Syntactic analysis musical texts. Novye informatsionnye tekhnologii v avtomatizirovannykh sistemakh. 16, 257–262 (2013) - in Russian
2. Yu, P.A., Danshina, M.V.: Methods for automatization of the process of decoding of znamenny chants. In: International Conference in Cognitive Linguistics, CrossLingua'2013: Cognition. Communication. Culture, Ukraine, Crimea (2013) - in Russian
3. Zelentsov, I.A., Philippovich, Yu.N.: Recognition of letters and words in ancient Russian cursive writing. Sci. Educ. 12, 27 (2011) - in Russian
4. Torgerson, W.S.: Theory and Methods of Scaling. Wiley, New York (1958). ISBN 0-89874-722-8
5. Specht, D.F.: Probabilistic neural networks. Neural Netw. 3, 109118 (1990)

Industry Papers

Enhancing Russian Wordnets
Using the Force of the Crowd

Dmitry Ustalov[1,2,3]($\boxtimes$)

[1] Krasovsky Institute of Mathematics and Mechanics, Ekaterinburg, Russia
[2] Ural Federal University, Ekaterinburg, Russia
[3] NLPub, Ekaterinburg, Russia
dau@imm.uran.ru

Abstract. The YARN (Yet Another RussNet) project aims at creating a large open machine-readable thesaurus for Russian using crowdsourcing. This paper describes the project itself along with its objectives and results of the pilot user study conducted at the end of 2013.

Keywords: WordNet · Crowdsourcing · Ontologies · Thesauri

1 Introduction

The YARN[1] (Yet Another RussNet) project was initiated in 2013 and aims at creating a large open WordNet-like machine-readable thesaurus for the Russian language using crowdsourcing techniques. The project objectives include creating the thesaurus, developing free and libré open source software to operate with it, designing the necessary data schemes and models, writing technical and user documentation.

Since there is still no open Russian thesaurus of acceptable terms of use, quality, and size available, the first milestone of the YARN project is the development of an online tool for assembling *noun synsets* based on the content of the dictionaries in YARN. The purpose of this stage is to create a linguistic and technical basis for further work, and also to evaluate the convenience of the project in the crowdsourcing aspect.

This paper is organized as follows. Section 2 focuses on related work in such areas as thesauri for several Slavic languages and Russian, thesaurus editing tools, and crowdsourcing approaches for ontologies. Section 3 briefly describes the philosophy of the YARN project and presents its design and implementation details. Section 4 analyzes the lessons learned after the pilot user study, which has been conducted in the Ural Federal University at the end of 2013. Section 5 concludes with final remarks and directions for the future work.

---

[1] http://russianword.net/

© Springer International Publishing Switzerland 2014
D.I. Ignatov et al. (Eds.): AIST 2014, CCIS 436, pp. 257–264, 2014.
DOI: 10.1007/978-3-319-12580-0_27

2   Related Work

Russian wordnets and thesauri. In fact, there is still no open Russian thesaurus of acceptable terms of use, quality, and size available for production-grade natural language processing and information retrieval applications including commercial ones.

- *RussNet* was launched in 1999 at the Saint-Petersburg State University [1]. The only publicly available project deliverable is a sample of 300 synsets constituted by verbs of emotional states[2].
- *RuThes* has been developed at the Moscow State University since 2002 and is intended to use in information retrieval applications [2]. A light version of this resource contains 55 000 concepts and is available[3] under the CC BY-NC-SA license.
- *RWN* was an attempt to translate Princeton WordNet into Russian by a group of researchers from the Novosibirsk State University [3] in 2003. The resulted dataset covers approximately 45 % of PWN entities and is freely available[4], but no systematic quality assessments of the obtained data were performed.
- *Russian Wordnet* was a collaboration of the Saint-Petersburg Transport University and the Russicon company to perform a semi-automatic translation of Princeton WordNet [4] in 2004. Unfortunately, project deliverables are unavailable for general public.
- *BabelNet* is a very large automatically generated multilingual thesaurus [5], the Russian part of which consists of 1.84M lemmas, 985K synsets, and 2.14M word senses[5]. No evaluation of the Russian data has been performed yet; the resource is released under the CC BY-NC-SA license.
- *Russian Wiktionary*[6] contains more than 158 689 word entries and 56 844 synonym relations as of March, 2012. The Wikokit project allows handling Wiktionary data as a relational database [6], but quality of such data is disputed.

Thesauri for Slavic languages. There are several wordnets for Slavic languages including Czech [7], Polish [8], and Ukrainian [9]. Since Slavic languages are highly inflectional and have a rich derivation system, in each case a special attention is paid to accounting for and encoding morphological characteristics.

Thesaurus editing tools. Examples of modern thesauri development tools are DEBVisDic [10], GernEdiT [11], as well as WordNetLoom [12] (see [12] for a brief overview of thesauri editing tools). Such ontology editors as protégé[7] and GATE Ontology Editor[8] have similar functionality. These tools are designed for professional lexicographers and are not intended to be used in crowdsourced projects when many untrained volunteers are involved to perform simple tasks in

---

[2] http://project.phil.spbu.ru/RussNet/
[3] http://www.labinform.ru/pub/ruthes/
[4] http://wordnet.ru/
[5] http://babelnet.org/stats.jsp
[6] http://ru.wiktionary.org/ and http://code.google.com/p/wikokit/
[7] http://protege.stanford.edu/
[8] http://gate.ac.uk/userguide/sec:ontologies:vr

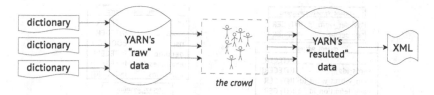

Fig. 1. The philosophy of YARN

quite short time. Wiktionary, OmegaWiki and related projects[9] are constructed with assumption that users are professional lexicographers familiar with the wiki markup.

Crowdsourcing approaches for ontologies. Amazon Mechanical Turk is de facto crowdsourcing tool for various linguistic resources including wordnets. Examples include ontological structure extraction from social tagging systems and engagement of turkers in the evaluation [13], creation of empirically-derived sense inventory [14], and of a semantic resource for production by solving three simple lexical substitution tasks [15], etc. Unfortunately, it is impossible to create a Russian thesaurus using MTurk since there are virtually no turkers from Russia present on the platform [16]. There are also other approaches available to bring crowdsourcing to ontologies with examples as modeling IDC-11 in WebProtégé [17], discovering correspondences between ontologies using CrowdMap [18] and linked data assessment using TripleCheckMate [19] (see [20] for more details).

3   YARN, A New Hope

The YARN project was initiated in 2013 and aims at creating a large WordNet-like Russian thesaurus using crowdsourcing (see the proposal paper [21] for more details). The fundamental difference between YARN and the above mentioned thesauri is in using the crowdsourcing techniques and previously developed dictionaries in order to create and enhance the thesaurus. YARN presupposes that (1) the native language of a crowdsourcing participant is Russian, (2) the participant has some understanding of lexicography basics, and (3) one has a mentor who can assist in case of trouble.

The philosophy of YARN is to take "raw" data from the existent dictionaries with appropriate licenses and let users organize their content into structured "resulted" data, see Fig. 1. Thus, one's objective is to refine "raw" synonymic rows and transform them into "resulted" synsets, which is probably easier and more convenient in contrast to doing this work manually or without "raw" data. Such a work is intended to be done using special user interfaces that incorporate YARN's "raw" data, display YARN's "resulted" data, and provides user a possibility to transform "raw" data into "resulted" data by performing simple operations.

---

[9] http://ru.wiktionary.org/ and http://www.omegawiki.org/

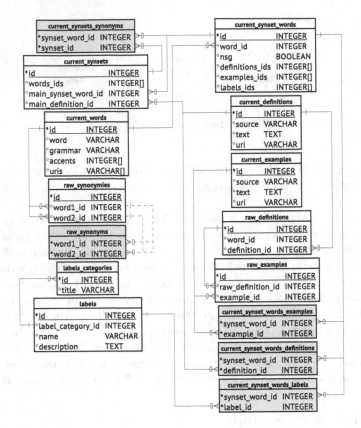

Fig. 2. The entity-relationship diagram of the YARN linguistic data (the highlighted tables are actually *views* in terms of relational databases)

Initially, YARN's "raw" data are composed of nouns, definitions, and examples from the parsed Russian Wiktionary [6] along with the noun lexical entries and definitions from the parsed Small Academic Dictionary, and noun synonymic rows from the parsed Abramov's Dictionary of Synonyms and Similar Expressions. Importing these dictionaries led to 67 962 noun entries having been added to the database; 21 659 of them being provided with frequencies based on the Russian National Corpus [22].

The brief structure of the YARN thesaurus is depicted at Fig. 2 excluding supplementary fields and history tracking tables. The core concept of YARN is *a synset*. A synset may contain *words*. The requirement of providing the word *definitions*, *examples* and *labels* makes it impossible to connect *a word* and *a synset* directly. Thus, there exists a supplementary entity named *synset_word* that links words and their definitions, examples and marks with the correspondent synsets.

From the technical point of view, the software is implemented using the Ruby on Rails framework and the PostgreSQL relational database with strong use of its specific features such as *array* data type, window functions, materialized views,

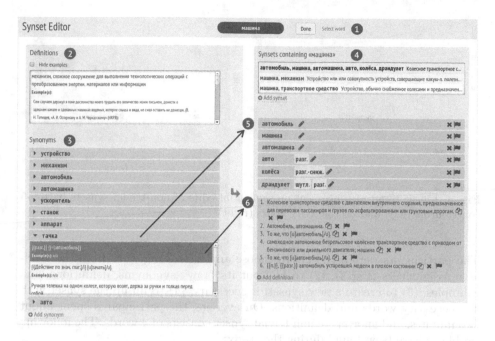

Fig. 3. An assignment choice dialog (captions are translated into English for convenience of readers; originally all interface elements are in Russian)

and full-text search. The YARN software exposes a well-documented JSON API endpoint[10] to make possible interaction with it. For instance, the synset assembly tool[11] shown at Fig. 3 is a single page JavaScript application that uses the JSON API to work with YARN. User authentication is performed through an OAuth endpoint provided by Facebook.

The YARN editor stores down the data modification history. In optimization purposes, several changes made by the same author within 12 hours are combined into a single change. Unfortunately, this prevents estimating the efforts and time span using only database records. However, it is possible to assign HTTP logs to synset revision history to achieve the full picture of the user activity. Therefore, two data sources were used during the user activity analysis: (1) depersonalized HTTP logs, (2) database records on synset modifications.

4   Pilot User Study and Results

During the recent pilot user study [23] in fall 2013, a group of 45 linguistics department sophomores and juniors of the Ural Federal University have assembled 970 *non-trivial*[12] synsets varying in quantitative and qualitative aspects.

---

[10] http://nlpub.ru/YARN/API
[11] http://russianword.net/editor
[12] A synset is called *non-trivial* if and only if it has more than one word.

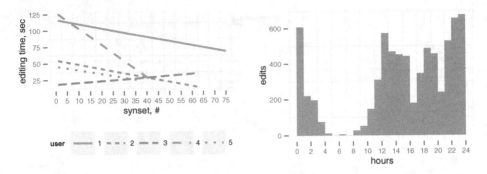

Fig. 4. Linear trend of time spent on sequential edits of non-trivial synsets by top-5 contributors (left) and user activity hours (right)

The participants were asked (1) to choose any interesting word from the available assignments (Fig. 3), (2) to find out its "raw" synonyms, definitions, and examples, (3) to create a new synset, if necessary, (4) to fill up the synset with the necessary words and definitions. On average, the participants spent about two minutes, and an average non-trivial synset consists of 4 words. The following problems have been found during the study:

- *Quality control.* It has become quite complicated to assess the data quality manually, e.g. spending only 3 min per synset will result in $3 \times 970 = 48.5$ hours of work. It is necessary to incorporate automated quality control mechanisms similar to [14].
- *Overdependence on "raw" data.* The study participants have found to believe that the dictionaries presented at the synset editing tool (Fig. 3) are the ultimate truth and contain no mistakes, noise, missing synonyms or definitions, etc. Such a behaviour may be avoided by both increasing the amount of the available "raw" data and introducing a content-based recommender system, which will assist a participant during his assignment.
- *Assignment generation.* At the moment of the study, all assignments were generated by descending order of the Russian National Corpus frequency. Since users are overdependent on "raw" data it is important to provide words for which there are as many relevant "raw" data available.

It is interesting that five of the most motivated participants have assembled the third of all non-trivial synsets (329 synsets). Four out of five most active participants demonstrate the learning effect as seen at Fig. 4 (left): the time spent reduces over each consequent synset. The participants were able to use the YARN synset assembly interface at any time of the day during the experiment. It is notable that the most of edits were done at the noon (12 a.m.), after supper (18 p.m.), and at the midnight, as it is seen at Fig. 4 (right).

5   Conclusion

The results of the conducted pilot user study demonstrate that YARN has a convenient and well-suited software to perform practical lexicographic work using crowdsourcing techniques.

The experiment has also showed that current drawbacks are caused either by weak sides of crowdsourcing, or by the interface shortcomings. Those can be solved by creating the supervisor-oriented interface, initial user testing on lexicography basics, publishing the interactive educational resources, etc.

Special attention should be paid at the technical aspects such as improving the user activity analysis tools and cleaning up the textual data from unsolicited markup.

The YARN software including the synset assembly tool is open source and is available on GitHub in a public repository[13]. The database is also available[14] in the XML format under the CC BY-SA license.

Acknowledgements. This work is supported by the Russian Foundation for the Humanities, project 13-04-12020 "New Open Electronic Thesaurus for Russian", and by the Ural Branch of RAS grant 12-S-1-1001/3. The author would like to thank the anonymous reviewers for their criticisms and suggestions.

References

1. Azarova, I., Mitrofanova, O., Sinopalnikova, A., Yavorskaya, M., Oparin, I.: Russ-Net: building a lexical database for the Russian language. In: Proceedings of Workshop on WordNet Structures and Standardisation, and How These Affect WordNet Applications and Evaluation, Gran Canaria, Spain, pp. 60–64 (2002)
2. Loukachevitch, N.V., Dobrov, B.V.: Development and Use of Thesaurus of Russian Language RuThes. In: Proceedings of Workshop on WordNet Structures and Standardisation, and How These Affect WordNet Applications and Evaluation, Gran Canaria, Spain, pp. 65–70 (2002)
3. Gelfenbein, I., Goncharuk, A., Lekhelt, V., Lipatov, A., Shilo, V.: Avtomatich-eskij perevod semanticheskoj seti WORDNET na russkij yazyk. In: Proceedings of Dialog-2003 (2003) (in Russian)
4. Balkova, V., Sukhonogov, A., Yablonsky, S.: Russian WordNet. In: Proceedings of the Second Global WordNet Conference, pp. 31–38 (2004)
5. Navigli, R., Ponzetto, S.P.: BabelNet: The automatic construction, evaluation and application of a wide-coverage multilingual semantic network. Artif. Intell. **193**, 217–250 (2012)
6. Krizhanovsky, A., Smirnov, A.: An approach to automated construction of a general-purpose lexical ontology based on Wiktionary. J. Comput. Syst. Sci. Int. **52**(2), 215–225 (2013)
7. Pala, K., Smrž, P.: Building Czech wordnet. Rom. J. Inf. Sci. Technol. **7**(1–2), 79–88 (2004)

---

[13] https://github.com/russianwordnet/yarn
[14] http://russianword.net/yarn.xml

8. Maziarz, M., Piasecki, M., Szpakowicz, S.: Approaching plWordNet 2.0. In: Proceedings of the 6th Global Wordnet Conference, Matsue, Japan (2012)
9. Anisimov, A., Marchenko, O., Nikonenko, A., Porkhun, E., Taranukha, V.: Ukrainian WordNet: creation and filling. In: Larsen, H.L., Martin-Bautista, M.J., Vila, M.A., Andreasen, T., Christiansen, H. (eds.) FQAS 2013. LNCS, vol. 8132, pp. 649–660. Springer, Heidelberg (2013)
10. Horák, A., Pala, K., Rambousek, A., Povolný, M.: DEBVisDic-first version of new client-server WordNet browsing and editing tool. In: Proceedings of the Third International WordNet Conference (GWC-06), pp. 325–328 (2006)
11. Henrich, V., Hinrichs, E.: GernEdiT: a graphical tool for germanet development. In: Proceedings of the ACL 2010 System Demonstrations, pp. 19–24 (2010)
12. Piasecki, M., Marcińczuk, M., Ramocki, R., Maziarz, M.: WordNetLoom: a WordNet development system integrating form-based and graph-based perspectives. Int. J. Data Min. Modell. Manage. 5(3), 210–232 (2013)
13. Lin, H., Davis, J.: Computational and crowdsourcing methods for extracting ontological structure from folksonomy. In: Aroyo, L., Antoniou, G., Hyvönen, E., ten Teije, A., Stuckenschmidt, H., Cabral, L., Tudorache, T. (eds.) ESWC 2010, Part II. LNCS, vol. 6089, pp. 472–477. Springer, Heidelberg (2010)
14. Rumshisky, A.: Crowdsourcing word sense definition. In: Proceedings of the 5th Linguistic Annotation Workshop, pp. 74–81 (2011)
15. Biemann, C.: Creating a system for lexical substitutions from scratch using crowdsourcing. Lang. Resour. Eval. 47(1), 97–122 (2013)
16. Ross, J., Irani, L., Silberman, M., Zaldivar, A., Tomlinson, B.: Who are the crowdworkers?: shifting demographics in mechanical turk. In: CHI'10 Extended Abstracts on Human Factors in Computing Systems, pp. 2863–2872 (2010)
17. Tudorache, T., Falconer, S., Noy, N.F., Nyulas, C., Üstün, T.B., Storey, M.-A., Musen, M.A.: Ontology development for the masses: creating ICD-11 in WebProtégé. In: Cimiano, P., Pinto, H.S. (eds.) EKAW 2010. LNCS, vol. 6317, pp. 74–89. Springer, Heidelberg (2010)
18. Noy, N.F., Sarasua, C., Simperl, E.: CROWDMAP: crowdsourcing ontology alignment with microtasks. In: Cudré-Mauroux, P., et al. (eds.) ISWC 2012, Part I. LNCS, vol. 7649, pp. 525–541. Springer, Heidelberg (2012)
19. Kontokostas, D., Zaveri, A., Auer, S., Lehmann, J.: TripleCheckMate: a tool for crowdsourcing the quality assessment of linked data. In: Klinov, P., Mouromtsev, D. (eds.) KESW 2013. CCIS, vol. 394, pp. 265–272. Springer, Heidelberg (2013)
20. Doan, A., Ramakrishnan, R., Halevy, A.Y.: Crowdsourcing systems on the World-Wide Web. Commun. ACM 54(4), 86–96 (2011)
21. Braslavski, P., et al.: YARN Begins. In: Proceedings of Dialog-2013 (2013) (in Russian)
22. Lyashevskaya, O., Sharov, S.: The Frequency Dictionary of Modern Russian Language. Azbukovnik, Moscow (2009)
23. Braslavski, P., Ustalov, D., Mukhin, M.: A spinning wheel for YARN: user interface for a crowdsourced thesaurus. In: Proceedings of the Demonstrations at the 14th Conference of the European Chapter of the Association for Computational Linguistics, Gothenburg, Sweden, pp. 101–104 (2014)

Search Engine of Mentions of Russian Companies on the Internet

Mikhail Khrushchev[✉]

SKB Kontur, Ekaterinburg, Russia
michael.khr@gmail.com

Abstract. This article describes a search engine of websites and mentions of Russian companies in the Internet. This system was developed by SKB Kontur company as a part of the product Kontur-Focus. The paper describes architecture of the system and Data Mining algorithms used in it.

Keywords: Search engine · Information retrieval · Data mining · Russian companies · Companies search

1 Introduction

Checking counterparties is a very important problem of modern business in the Russian Federation. To solve this problem CJSC SKB Kontur has developed a web service Kontur-Focus, which contains open databases, such as the Unified State Register of Legal Entities of the Russian Federation, the Unified State Register of Private Entrepreneurs of the Russian Federation, the Register of arbitration cases. It also contains some analytic data about companies and a search engine that allows users to get information about the companies they are interested in. However, this data is not often enough to compile a complete picture of a company. Moreover, the information from such open data sources is not up-to-date.

A large number of companies have their own websites and pages in open catalogs of organizations. Besides, it is also possible to find information about companies on different websites. Companies are interested in revealing current data or up-to-date information about them on their web pages. By using information from the World Wide Web it is often possible to determine a sphere of company's activity and to evaluate a company as a counterparty. So, it makes sense to show users those websites where the company, they are interested in, is mentioned. To solve this problem, we developed a search engine of mentions of organisations in the Internet.

In this work we describe architecture of the system and its main components. Also, much attention is paid to the problem of binding companies with data about companies from the web pages. This problem has much in common with Record Linkage problem [1,2] and it is solved using the probabilistic approach.

D.I. Ignatov et al. (Eds.): AIST 2014, CCIS 436, pp. 265–273, 2014.
DOI: 10.1007/978-3-319-12580-0_28

2   Problem Statement

The web service is a database of all organizations and private entrepreneurs of the Russian Federation. For each company a database may comprise of the following fields: a full name, a short name, outdated names, phones, addresses, full names of executives and founders, a PSRN (primary state registration number), a TIN (taxpayer identification number), names and company details of the founders. For each private entrepreneur only a full name, PSRN and TIN are known. Some of this data except PSRN and TIN may be missing, distorted or not reflect the reality.

So, it is necessary to create a binding, i.e. to indicate for each company as many websites and web pages with company's mentions as possible. In other words, the system must build a set of pairs (company, website).

It is important to note that a company may be bound to a huge number of web pages. So it makes sense to range list of websites according to user's interest so that company's website goes first.

3   Assumptions

At the moment of writing this paper a number of second-level domains in zones .ru and .su exceeded 6 million domains. Website structures are now vary a lot; each new website may have a unique structure. So it is necessary to make some assumptions about structure of websites:

1. A structure of any website is a tree with web pages as vertices;
2. All nodes of each tree are always from single second-level domain;
3. The root is always a web page with an empty URL path;
4. If this URL redirects somewhere else, then there is a node for this URL with a single child - a web page where the redirect points to;
5. A website may be presented as a graph whose vertices correspond to the pages, and edges correspond to the links from one page to another. Our tree coincides to one of the trees constructed by breadth first search (BFS) with root being a starting vertex.
6. If the company is mentioned on some website, then there is some subtree of this website that is related to the company. So, if the website is the company's website, then the required subtree is an entire tree of the website. If the company is mentioned in a catalog of organizations, then the required subtree is the company's page in that catalog.

4   Overall Architecture

Loading page and its analysis are done by the following scheme (see Fig. 1):

1. The system downloads web pages from the Internet in a certain order.
2. From each web page the system extracts special elements (features). Each feature may correspond to name, phone, address, PSRN and TIN of a company or a private entrepreneur. Also it may correspond to the full names of people that may be connected with the companies.
3. The system finds the companies-candidates that correspond to some features from the set of features that was built at the previous step.
4. For each second-level domain trees of web pages from this domain are built. The system determines in which subtrees which companies are mentioned.
5. When the user sends a query to the web service to get information about an organization, the system builds an ordered list of sites bound to the organization and sends it back.

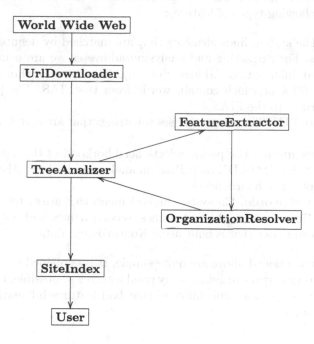

Fig. 1. The system architecture

4.1 Web Pages Downloading

For getting data from the Internet we are using the Apache Nutch service[1] [7]. The Apache Tika service is applied for processing all acquired data. Downloading starts with root pages of second-level domains of first-level domains .ru and .su. The system downloads and analyzes the web page only if a link to this web page from downloaded web pages (outlink) exists and at least one of the next two conditions is true:

---

[1] http://nutch.apache.org/

1. The URL of the web page is matched by a template from a defined set of templates. The set contains such words as "contacts", "about" and other words that may correspond to the web pages with details about a company.
2. Description of the outlink (anchor) is matched by a template from a defined set of templates such as "our contacts", "about company" and etc.

These rules allow to avoid downloading and analyzing those pages that do not contain company's details and mentions.

4.2 Feature Extracting

The next step is extracting features, i.e., elements of pages that correspond to companies details and full names of people. During this procedure the system extracts the following types of features:

1. **Phones.** The system finds elements that are matched by defined templates.
2. **Addresses.** For extracting and analysing addresses we are using the FIAS (the Federal Information Address System). Parsing algorithm selects such substrings, 50 % of which contain words from the FIAS. The parser maps such substrings to the FIAS.
3. **PSRN and TIN.** The parser chooses substrings that are matched by defined templates.
4. **Company's name.** The parser selects neighborhood of the key words such as "JSC", "CJSC", "LTD", their Russian analogs and others. Also, the parser chooses quotes neighborhood.
5. **Full names of people.** The system finds elements that are matched by defined templates. To determine the first names, second names and patronymic the parser uses statistics that is built using Kontur-Focus data.

Templates mentioned above are quite simple. It is explained by the fact that at this step it is important to get as many good features as possible. Furthermore, due to the next steps, the probability of that bad features influencing the final result is very small.

4.3 Address Parsing

To compare addresses it will be convenient to map them to one system. For this problem we are using the FIAS. One can represent the FIAS as a tree with such conditions:

1. The root of the tree is a node that corresponds to the Russian Federation
2. One node is a descendant to another if the first entirely belongs to the second as a geographic object. So, each address presented in the form of a text may correspond to some branch of the FIAS tree.

Let us assume that a branch b of the tree T contains a word w if at least one of the nodes of b corresponds to the Russian Federation member, the name of which contains the word w. The word w belongs to a text address $addr$ if $addr$ contains w as a word of the Russian or English language.

Then assume that the text address $addr$ corresponds to the branch b_{addr} if the next condition is true:

$$b_{addr} = \underset{b \in T}{\operatorname{argmax}} \sum_{w \in b} \operatorname{ComLen}(w, addr), \tag{1}$$

$$\operatorname{ComLen}(w, addr) = \begin{cases} 1, & \text{if } w \in addr \\ |w|, & \text{if } w \notin addr \end{cases}, \tag{2}$$

where $|w|$ is a word length.

Such presentation is convenient to use when comparing the address for equivalence or nesting. However, this model doesn't take into account the numbers of houses and offices, because this type of data is not well-presented in the FIAS. One can find that this model considers the same words from an address only one time. Because of that some information about an address may be lost. For example, while parsing "Moscow, Moscow street" we lose information about the street because the name of the street and the city are the same. Due to the nature of the Russian language, the number of such addresses is very small because of the suffixes. However, such addresses may be seen in Kontur-Focus.

4.4 Searching Mentions of Companies on Web Pages

At this stage the system has to find the companies that may be mentioned on a selected web page by using a set of features from this web page. To achieve that it will make sense to build some estimation – it will estimate how close the selected company is to the set of features from this web page. For such a metric it will be helpful to use the likelihood that a randomly taken company from Kontur-Focus has the same set of features as our company in this estimation:

$$P_{org}(names, addrs, phones, tins, psrns, persons), \tag{3}$$

where $names, addrs, phones, tins, psrns, persons$ are the features found on the web page.

It is convenient to assume that probabilities of appearing features of different types are independent. But it makes sense to consider dependence of the city's phone code and the city from the address. These parameters are really dependent. Ignoring this dependence may lead to the errors of the model further on. So, the likelihood estimation may be defined as follows:

$$P_{org}(names, addrs, phones, tins, psrns, persons) =$$
$$\prod_{feat \in feats \backslash addrs} P_{org}(feat) \prod_{addr \in addrs} P_{org}(addr|phones), \tag{4}$$

where

$$feats = \{names \vee addrs \vee phones \vee tins \vee psrns \vee persons\} \tag{5}$$

$$P_{org}(feat) = \begin{cases} P(feat), & feat \in org \\ 1, & feat \notin org \end{cases} \tag{6}$$

$$P(addr|phones) = \max_{phone \in phones} \frac{P(addr)}{P(city_{addr})} \frac{P(city_{addr} \wedge code_{phone})}{P(code_{phone})} \tag{7}$$

where we assume that feature $feat$ belongs to the company org if it corresponds to one of the company's fields.

For each feature $feat$ the system builds $P(feat)$ by using information about the amount of fields in the Kontur-Focus that corresponds to $feat$ and information about the probability that the feature $feat$ appeared mistakenly.

$$P(feat) = P_{stat}(feat)(1 - P_{error}(feat)) + P_{error}(feat) \tag{8}$$

For names of a company probability is calculated from the frequency of appearing in the database of Kontur-Focus coincident words, bigrams and complete names between company and features.

The P_{org} function is modelling the probability of the fact that a randomly taken company from Kontur-Focus has the same set of features as our company. This function is proportional to the amount of organizations with the same set of features and it is convenient to use it to estimate reliability of accordance of a company and the found features. Hence, we can use it to estimate the quality of binding between a company and a web page.

This method has much in common with Naive Bayes classifier [3,4]. It also uses information about dependencies and error probabilities.

Searching for a company is made according to the following scheme:

1. For each feature the system finds a list of candidates – the companies with a field that corresponds to the given feature. Those features, for which too many companies are found are removed from consideration.
2. For each found company P_{org} is calculated.
3. A company is removed from consideration if P_{org} is greater than some threshold. This threshold should not be too strict. We use it to reduce the amount of considered companies for optimisation purposes but not to select the companies that are mentioned on this page with guaranty. Analysis of other pages may improve this estimation later.

4.5 Building of Bindings Between Companies and Subtrees of Websites

For each second-level domain construction of bindings is done using the following scheme:

1. On all pages of the selected domain the system builds website trees. Building is generated by the next principle:
   (a) The algorithm orders URL by amount of chars '/' in the path and by domain level.
   (b) On each next step the algorithm selects the page which has not yet been used. Then it builds the tree with the root in this node by BFS.
2. For each tree the system builds a set of bindings between companies and subtrees of the website.

For each feature type $Feat$ that may be an address, a phone, a company name, a PSRN, a TIN or a person name, we declare function $P_{(org,Feat)}(N)$ that equals to $P_{org}(feats)$ where $feats$ – features of type $Feat$ that was extracted from the web page N. Then introduce functions $P(org, Feat)(T_N)$ and $P(org)(TN)$ where T_N is a node corresponding to the web page N:

$$P_{(org,Feat)}(T_N) = \min(\min_{C \in Children(T_N)}(P_{(org,Feat)}(C)), P_{(org,Fcat)}(N)) \qquad (9)$$

$$P_{(org)}(T_N) = \prod_{feat \in \{addr,phone,name,person,psrn,tin\}} P_{(org,feat)}(T_N) \qquad (10)$$

In this approach we consider summary information about features from different pages of a website. For example, if there is a phone number on the web page "Our contacts" and there is a full name of the general manager on the web page "Management" then algorithm will use both of these factors while binding. There is one additional condition: the algorithm binds a company org to a node T_N only if $P_{(org,feat)}(N)$ is small enough for each feature of type $feat$. In other words, the algorithm must be "sure" that company org is mentioned on web page N to build this binding. This condition allows to stop in the root of the subtree, that is likely to characterize the company.

In addition, it will be convenient to use information outside the tree of the website while binding. Often, links to a company's website may be found on a different website in the context of this organization. In other words, there is a high probability that a page with a link to a company's website contains details of that company or the full names of its managers or founders. We call such pages inlinks. Inlinks may contain a large amount of useful data, and we also can use it while binding.

4.6 Ranking of Bindings

When a user sees a list of websites about a company, it is very important to show him bindings for this company in the right order. So, it is necessary to build a ranking function to order bindings. Page ranking is a well known problem and there are many different solutions [5,6] to it. Such methods attempt to predict popularity of the page to give them higher rank. However, often most interesting website for the user – the website of the company is less popular than a website that contains mentions of that company. So, we try to predict popularity of page in the context of the company.

The most interesting websites are websites of the companies. Based on this fact and some other factors we developed a ranking function. This function uses information about the amount of website pages which are bound to this company, information about the amount of other companies that are bound to the website and information about the number of inlinks. The latter helps to identify whether the given website is the company's and how often it is mentioned in the context of the given organization.

5   Evaluation

To check quality of the bindings algorithm, we used a database of the company 2GIS. This database was obtained using 2GIS API with agreement of the service developers. It was obtained only with the purpose of research.

For each organization 2GIS' API returns data that contains the address, name of the company and the company's website. By using this data we built 2GIS bindings: each company from 2GIS we try to bind to some company from the Kontur-Focus by name and address. These bindings perform well but the quantity is not great. If we assume that the bindings built by 2GIS and bindings built by our algorithms are independent then it will be possible to estimate the percentage of found correct bindings to all possible correct bindings. The companies often provide the 2GIS with information about their main website. So we can use this information to estimate quality of ranking function. The statistic of comparing bindings:

- Our bindings contain about 65.6 % of the 2GIS bindings;
- Sites from 57.7 % of the 2GIS bindings(88 % of bindings from previous block) take the first place after ranking.

It will be convenient to assume that the number from the first block is *recall* of the system. It is not completely fair, because the goal of our algorithms is to search mentions of the companies but not only companies' websites. However, it is hard even to estimate an amount of all mentions in the Internet. Also, search of companies' websites is one of the main tasks of our algorithms.

We will say that a binding is correct if the web page of this binding contains real mention of the company of the binding. To calculate *precision* we manually checked about 500 random bindings that were built by our system. We found out that 57.6 % of our bindings are correct.

6   Conclusion

In this paper we presented a mechanism for building bindings between companies from Kontur-Focus and websites from the Internet. Its work was evaluated and the evaluation results are follows:

- Recall of the system is about 65.6 % on 2GIS data.
- Precision of the system is about 57.6 %.

This quality was considered sufficient and at the moment the service that works using these mechanisms has been developed and implemented. The service is available at the website of Kontur-Focus system.[2] In the future, it is planned to improve bindings quality by analysing publicly available catalogs of organizations and improving algorithms.

Acknowledgements. We would like to express our gratitude to the 2GIS team that collected and provided the data helped us estimate quality of our approach. Also, we would like to express our gratitude to Vladimir Gusev from Kontur Labs, who helped us with creation of 2GIS bindings. Last but not least we would like to thank the anonymous reviewers and the volume editors.

References

1. Christen, P.: Data Matching. Springer, Heidelberg (2012)
2. Scheuren, F.J., Herzog, T.N., Winkler, W.E.: Data Quality and Record Linkage Techniques. Springer Science+Business Media LLC, New York (2007)
3. Irina, R.: An empirical study of the naive Bayes classifier. In: IJCAI 2001 Workshop on Empirical Methods in Artificial Intelligence (2001)
4. Murphy, K.P.: Naive Bayes Classiers. University of British Columbia, Canada (2006)
5. Page, L., Brin, S., Motwani, R., Winograd, T.: The PageRank citation ranking: Bringing order to the web (1999)
6. Selvan, M.P., Chandra Sekar, A., Priya Dharshini, A.: Survey on web page ranking algorithms. Int. J. Comput. Appl. **41**(19), 1–7 (2012)
7. White, T.: Hadoop. The Definitive Guide, 2nd edn. OReilly Media, Inc., Massachusetts (2010)

---

[2] https://focus.kontur.ru/

Untangling the Semantic Web: Microdata Use in Russian Video Content Delivery Sites

Andrey Kutuzov[✉] and Maxim Ionov

Mail.ru Group, Moscow, Russia
{andrey.kutuzov,m.ionov}@corp.mail.ru

Abstract. Nowadays, more and more sites incorporate semantic markup into their pages. This allows search engines to better understand the content of the webpage. This paper investigates the deployment of semantic markup in the form of Microdata on Russian video content delivery sites. We point out commonalities and common problems and link our data set to DBpedia. General description of quantitative and qualitative features of semantic markup usage is given, based on large dataset crawled from Russian Internet segment.

Keywords: Semantic web · Information retrieval · Microdata · RDF

1 Introduction

More than 20 years ago Tim Berners-Lee envisioned the Web which would possess meaning, a web of data that can be processed by machines. This was seen as a basis for much better coherence and much more efficient information retrieval on global scale. However, only recently World Wide Web started showing signs of wide adoption of these ideas. More and more sites start deploying semantic markup on their pages.

The present paper aims to describe the deployment of semantic markup in the form of Microdata[1] in one part of Russian segment of the Internet, namely video content delivery sites and video hosting services. We point main highlights, quantitative features and common problems, and attempt to link this set of structured content to other global data storages.

2 Related Work

Research into semantic markup deployment over general Internet started only recently. Bizer et al. in [1] used web corpus published by Common Crawl foundation to describe how various flavours of Semantic Web are integrated into web pages globally. Percentage of sites using Microformats, RDFa and Microdata is given throughout the world, along with most frequent classes and properties.

---

[1] http://www.whatwg.org/specs/web-apps/current-work/multipage/microdata.html

© Springer International Publishing Switzerland 2014
D.I. Ignatov et al. (Eds.): AIST 2014, CCIS 436, pp. 274–279, 2014.
DOI: 10.1007/978-3-319-12580-0_29

The authors conclude that semantic markup has found considerable adoption on the Web.

More particular research about semantic markup deployment on national level is given in [2]. The paper overviews the use of Semantic Web technologies in Austrian commercial sites. Interestingly, findings of Loibl in [2] and Bizer et al. in [1] about the prevalence of microformats or RDFa differ. Bizer et al. argue that microformats prevail over RDFa over all the Web, while [2] observes dominance of RDFa. The present paper among other issues addresses this one in the context of Russian movie sites.

We are not aware of any published academic work analyzing deployment of semantic markup over Russian Internet in detail.

3 Semantic Markup Deployment in Russian Video Content Sites

3.1 Source of Data

24 % of web pages in Russian Internet segment carry semantic markup in some form[2]. One important type of semantic markup adopters are video content delivery sites. Generally these resources provide user with the possibility to watch video content online without downloading it.

User intent which these sites answer is quite popular. Nearly 10 % of daily 40 million queries processed by Mail.ru search engine are of this type: a user seeking some particular movie or short clip to watch. 23 of top 50 (and 6 of top 10) sites most frequently appearing on search engine results page are video hosting services.

Close to 50 % of top Russian video content delivery sites deployed semantic markup. Accordingly, both major Russian commercial search engines (Yandex and Mail.ru) routinely use semantic meta data inside these sites' pages to better process their content.

Our research was performed within Mail.ru search engine and we studied the copy of Russian Internet made by Mail.ru crawler. Among others, it regularly downloads the content of several most popular sites distributing video content and employing semantic markup. Popularity is judged by the frequency of the site appearing in search results and its click rank. The list is as follows: actorpedia.net, amazingcinema.ru, baskino.com, bigcinema.tv, gos-kino.ru, ivi.ru, kinopoisk.ru, kiniska.com, kinoestet.ru, kinomatrix.com, kinoprosmotr.net, kinostok.tv, megogo.net, multiki-online.net, mult-online.ru, newstube.ru, ovideo.ru, ruseriali.com, rutube.ru, smotri.com, youtube.com, zerx.ru

Altogether this makes for approximately 100 million URLs (the exact number increases with each crawling session). The majority of URLs naturally come from *youtube.com* (about 83 million pages), and its Russian analogues *rutube.ru* (2.2 million pages) and *smotri.com* (about 1.2 million pages). All possible semantic-markup is extracted from the pages of these sites and used as structured data to

---

[2] Yandex estimation, http://tech.yandex.ru/events/yagosti/wsd-msk-dec-2013/talks/1517/.

construct better snippets (including movie description, its duration, actors' and producer name, etc.).

We filtered out data extracted from sites with user generated video content (like *youtube.com* and its Russian analogues) and remained with 18 sites delivering movies and TV series. All of them are Russian-based and deliver content mainly in Russian. In total the set consisted of about 1.5 million web pages with semantic markup.

3.2   Types of Markup Used

Standards used for markup differ. But most informative of all was found to be Microdata with Schema.org vocabulary. Many pages come with one and the same information deployed in several standards. However, almost always Microdata provided the most extensive meta data about the movie in question.

In [2] the author argues that RDFa semantic formats are dominant in Austrian commercial sites. We checked our data set and found that, in accordance with Yandex estimation (see above), Opengraph protocol[3] of RDFa family is also used extensively on the pages in our collection. However, in the majority of cases properties set is quite limited and does not exceed a trivial triple "title, URL, description". That is quite insufficient to describe a movie. It is possible that this is also the case with the sites in [2]. At the same time, as stated in [1] on account of general Web, 'the number of Microdata properties used is about twice as large as the number of RDFa properties indicating that Microdata annotations are on average more fine grained than RDFa annotations'. This correlates with our evidence. It is interesting that Bizer et al. in [1] do not list http:// schema.org/Movie class among frequent Schema.org classes and do not study its properties distribution. However, their conclusions remain true for our data set. The only difference to observations in [1] was that in our case microformats (classes like hCard, hCalendar and others) were not found to carry movie-related information in a more or less convincing frequency.

Below is a typical example of a page with Microdata deployed (from *ivi.ru*, only meaningful excerpts are shown, Russian text translated into English):

```
<div class="content-main" itemscope itemtype="http://schema.org/Movie">
<meta itemprop="name" content="Live broadcast"/>
<div itemprop="description"> <p>Soviet social drama //
<Live broadcast> tells the story of old friends...</p> </div>
```

Further we study a collection of statements about things, which is the essence of RDF ideology, where semantic descriptions consist of statements, properties and classes. Linguistically, the set consists of triples: subjects, predicates and objects. 'Things' in our case are movies, their properties ('predicates') are listed in <http://schema.org/Movie> vocabulary and the values ('objects') of these predicates differ for each movie in particular.

Here is the data from the example above serialized with n3:

---

[3] http://ogp.me/

```
_:node4c1cbf2961c80cb130864d6b14da36 a <http://schema.org/Movie> ;
<http://schema.org/Movie/description> "Soviet social drama..." ;
<http://schema.org/Movie/name> "Live broadcast" ;
<http://schema.org/Movie/director> _:nodea9c4fdb63742741da3d3895b97d19338 .
_:nodea9c4fdb63742741da3d3895b97d19338 a <http://schema.org/Person> ;
<http://schema.org/Person/name> "Oleg Safaraliev" .
```

One can see how n3 here describes two entities, each marked with unique blank node (starting with an underscore). Note that no site from those we studied has implemented linking with other databases (like DBpedia or Linked Movie Database) yet. We address this issue in the Sect. 4.

In total our data set contains nearly 1.13 million entities belonging to the type <http://schema.org/Movie>. As expected, a lot of them are duplicates. Counting unique titles only, we get about 130 000 separate movies. About 70 % of entities come from *kinopoisk.ru*, 10 % from *kinoestet.ru*, 7 % from *ivi.ru*, and 4 % from *ovideo.ru*. *kinomatrix.com* and *baskino.com* together contribute to 2 % of the entities, other sites are even less.

Below we give some statistics for Microdata deployment over Russian movie sites.

3.3 Distribution of Predicates and Objects

Most common properties among entities of <http://schema.org/Movie> type in our data set are *name, image, genre* and *duration. alternativeHeadline* property is usually used to tag English title of the movie. *director, actors, contentRating, producer, musicBy, description* and *aggregateRating* can also be met in more than 60 per cent of resources. Below these twelve properties frequency drops dramatically and other predicates can be safely considered to be uncommon in the wild. One can note that standard-recommended predicate <http://schema. org/actor> can be met only in one resource in a hundred, while legacy <http:// schema.org/actors> is as common as 83 %.

We also performed analysis of the distribution of number of properties for movie entities (how many different predicates are used). Typical number of unique properties is 12 (62 % of all entities) and they are as follows: *genre, name, actors, description, producer, duration, alternativeHeadline, musicBy, aggregateRating, image, contentRating, director*. Entities with more than 13 predicates usually add *productionCompany, dateCreated, datePublished* and *inLanguage* properties.

Most of entities with a number of predicates equal to 12 or 10 come from *kinopoisk.ru* or *ivi.ru* sites. Notably frequent entities with 4 predicates (6 % of all entities) usually come from *kinoestet.ru* and contain only *name, url, image* and *ratingValue* properties. As for genre diversity, it follows long-tailed distribution with 'drama' being the most popular movie genre.

We also analyzed how often persons (directors and actors) are represented as literals and as proper entities (separate blank nodes). As stated earlier, Semantic Web approach presupposes that if anything has a corresponding class in the vocabulary than it should be described as an entity of this class (in this case,

<http://schema.org/Person>. However, often webmasters choose to simply put a literal string as a value of <http://schema.org/Person/actor> predicate and get away with that. This strongly undermines the whole idea of Web as data, as it makes impossible to link to this particular person or assign additional properties to it as an entity.

4   Linking the Set to External Databases

In order to enrich our database we performed linking with DBPedia [3]. This process consisted of two steps: mapping names from our database to target database and checking mapping on other fields, like *director* and *actor*. In case of database interlinking precision is more important than recall (we wanted to get less but more precise information) hence it was important to properly check mapping hypotheses. These steps were performed on a subset of the database consisting of 12 962 titles.

On the mapping stage for each movie title from our database we requested <http://dbpedia.org/ontology/Film> entities with rdf:label matching to the title. This way we extracted 3 350 entities ($\approx$25.8 % of all titles). The reason for low recall is absence of an exact label of type Film on DBPedia. The most frequent case is additional words in title for disambiguation. Movies usually has string '(film)' after title. By simply adding this string to the title we were able to increase recall by about 50 % (from 18.2 % to 25.8 %).

After this step we performed checking to determine if extracted connections were right. We tried to match frequent fields: *director, producer, alternativeHeadline* and *musicBy*, but we chose to do matching using only *alternativeHeadline* and *director* fields, because they are the only fields appearing in more than 50 % entities.

Naïve matching using exact string comparison gave us 2 303 matches ($\approx$68.7 %) by matching *alternativeHeadline* and only 31 matches ($\approx$0.9 %) by matching *director*. False negatives can be divided into three main classes: (1) *alternativeHeadline* contains English movie title rather than original one. (2) Some movie titles and directors names contain orthographic differences. To overcome this problem we changed exact string comparison to fuzzy string matching using Levenstein's distance and allowing one mistake. (3) Director's name is organized differently in DBPedia and our database. So we matched first and last names without taking order into consideration. After these tweaks we got 2 393 ($\approx$71.4 %) title matches, 1 477 ($\approx$44.1 %) director's name matches and 2 672 ($\approx$79.8 %) total matches counting results with at least one match.

We took small samples from four groups and checked them manually: (1) matches by both fields (25 matches, 0 mismatches), (2) matches by field *alternateHeading* and mismatches by field *director* (21 matches, 4 mismatches), (3) mismatches by field *alternateHeading* and matches by field *director* (25 matches, 0 mismatches), (4) mismatches by both fields (9 matches, 16 mismatches). Results show that we can link with confidence checking only one field, however this will lead to a great amount of false negatives, which can be decreased by accepting every link that has at least one field matching. False mismatches in *director* fields appear from different orthography and synonyms (e.g. "brothers Coen" vs. "Ethan Coen, Joel Coen"). False mismatches in *alternateHeading* are mostly caused by titles in languages different than English.

To increase recall and precision one should add more fields to the matching process and use different databases like LinkedMDB[4] to increase amount of data to check. This is a subject for future research.

5   Conclusion and Future Work

We described *status quo* of semantic markup deployment in the video content delivery sector of Russian segment of the Internet. We conclude that all types of markup are extensively used in this field, but Microdata with Schema.org vocabulary is the most informative in most cases.

General account on Microdata usage was given and several important webmaster mistakes highlighted. We also linked entities (movies) from our data set to DBpedia resources where possible. The dataset on which our paper is based is available online under Creative Commons Attribution-ShareAlike license at http://ling.go.mail.ru/semanticweb/. Our first step in further work will be to analyze this data deeper and provide more statistics to better understand how webmasters use semantic standards. This will demand increasing the number of sites under scrutiny. Also, there is an ongoing work to integrate Semantic Web principles into Mail.ru search engine. We plan to move from simply injecting RDF data into search snippets to studying the possibility to use inferential similarity between video entities to improve search results.

References

1. Bizer, C., Eckert, K., Meusel, R., Mühleisen, H., Schuhmacher, M., Völker, J.: Deployment of RDFa, microdata, and microformats on the web – a quantitative analysis. In: Alani, H., et al. (eds.) ISWC 2013, Part II. LNCS, vol. 8219, pp. 17–32. Springer, Heidelberg (2013)
2. Loibl, W.: Where is the semantic web? an overview of the use of embeddable semantics in Austria
3. Lehmann, J., Isele, R., Jakob, M., Jentzsch, A., Kontokostas, D., Mendes, P.N., Hellmann, S., Morsey, M., van Kleef, P., Auer, S., Bizer, C.: DBpedia - a large-scale, multilingual knowledge base extracted from wikipedia. Semant. Web J. **26**, 1–58 (2014)

---

[4] http://www.linkedmdb.org/

Author Index